ELASTOMERS:
CRITERIA FOR ENGINEERING DESIGN

Dr A. R. PAYNE, M.Sc., D.Sc., F.Inst.P., C.Eng.
M.I.Mech.E., M.B.I.M., F.P.R.I., F.R.S.A.

ELASTOMERS:
CRITERIA FOR
ENGINEERING DESIGN

Edited by

C. HEPBURN

Senior Lecturer in Rubber Technology,
Loughborough University of Technology, UK

and

R. J. W. REYNOLDS

Professor Emeritus, Loughborough University
of Technology, UK

APPLIED SCIENCE PUBLISHERS LTD
LONDON

APPLIED SCIENCE PUBLISHERS LTD
RIPPLE ROAD, BARKING, ESSEX, ENGLAND

British Library Cataloguing in Publication Data

Elastomers.
  1. Elastomers—Addresses, essays, lectures
  2. Rubber—Addresses, essays, lectures
  I. Hepburn, C   II. Reynolds, Reginald John
  William   III. Payne, Arthur Robert
  678   TS1927

  ISBN 0-85334-809-X

WITH 35 TABLES AND 227 ILLUSTRATIONS

© APPLIED SCIENCE PUBLISHERS LTD 1979

Printed in Great Britain by Galliard (Printers) Limited, Great Yarmouth

FOREWORD

It is very seldom that a group of scientific organisations combines together to stage a symposium in memory of a distinguished man and in the subject of their interests. Bob Payne contributed enormously to the science and technology of elastomers. The variety of behaviour and the rational explanation of the phenomena are probably greater than in other non-metallic materials. In spite of their scientific and technological importance, there seemed in the past to be almost an unwillingness among scientists to attempt to unravel these phenomena, many of which have been long observed in technological practice.

Bob Payne was one of the small band who clearly realised what had to be done and, what is more, set about doing it in the various organisations in which he worked. He covered the whole field from the fundamental to the immediately practical. To make progress in the latter it was necessary to add to the basic knowledge of the former. The papers given at the symposium form a fitting and just tribute to a man whose contributions were so outstanding.

Sir Harry W. Melville, KCB, FRS

PREFACE

As a tribute to Dr A. R. (Bob) Payne, who died in August 1976, several organisations with whom he was closely associated during his career thought it appropriate to hold a Symposium—*Elastomers: Criteria for Engineering Design*—in his memory promoted jointly by Loughborough University of Technology, PRI, MRPRA, RAPRA and SATRA. The Symposium, held at Loughborough University of Technology in April, 1978, was concerned with the academic, theoretical and practical aspects of rubber physics and engineering, ranging from the basic physics of elastomers to their industrial applications. Papers from many well-known authorities were presented and it was decided to publish these papers as a commemorative volume.

This wide subject was of deep interest to Dr Payne and he is acknowledged to have made a significant contribution in many aspects. Important testing devices he developed included the curometer, which follows the changes in physical properties in rubber while it is curing, a sinusoidal strain dynamic tester, and a device for studying the dynamic properties of rubber, particularly the dispersion of carbon black during processing. He also made a major contribution to understanding the strength and reinforcement of filled rubbers, particularly as regards effects of hysteresis on these properties. He had three books, and well over one-hundred papers on rubber physics and engineering, published during his lifetime.

His professional appointments have been as Principal Scientific Officer of RAPRA (1952–1962), Principal Physicist of MRPRA (1962–1968), and Director of SATRA (1968–1976). He was appointed Managing Director of RAPRA and should have taken up the position in September 1976.

Dr Payne was awarded the Colwyn Medal in 1972 by the Institution of the Rubber Industry. In the same year he became an Industrial Professor of the Institute of Polymer Technology, Loughborough. His other significant external professional positions included being chairman of the IRI (1970–72) and chairman of CDRA–Committee of Directors of Research Associations (1975–76).

The editors wish to express their thanks to all who have made this book possible: to authors for allowing their papers to be published in this form, to Mr Iain Howie and his colleagues at SATRA for invaluable assistance in preparation of the scripts, to Loughborough University of Technology for facilities and to the publishers for their understanding and support.

C. HEPBURN
R. J. W. REYNOLDS

LIST OF CONTRIBUTORS

K. W. ALLEN
Lecturer in Adhesion Science, Department of Chemistry, City University, London EC1V 4PB, UK.

S. M. DEAN
Research Worker, Department of Chemistry, City University, London EC1V 4PB, UK.

J. B. DONNET
President of the University of the Upper Rhine, 68200 Mulhouse, France.

A. N. GENT
Professor of Polymer Physics and Assistant Director, Institute of Polymer Science, University of Akron, Ohio 44325, USA.

P. GROOTENHUIS
Professor in the Department of Mechanical Engineering, Imperial College of Science and Technology, London SW1 2BX, UK.

D. C. HARGET
Technical Representative, Elastomers Department, DuPont (UK) Ltd, Maylands Avenue, Hemel Hempstead, Hertfordshire HP2 7DP, UK.

C. HEPBURN
Senior Lecturer, Institute of Polymer Technology, Loughborough University of Technology, Leicestershire LE11 3TU, UK.

B. P. HOLOWNIA
Lecturer, Department of Mechanical Engineering, Loughborough University of Technology, Leicestershire LE11 3TU, UK.

P. G. HOWGATE
Head of Plant and Dynamic Engineering, Rubber and Plastics Research Association, Shawbury, Shropshire SY4 4NR, UK.

A. KADIR
Research Worker, The Malaysian Rubber Producers' Research Association, Brickendonbury, Hertfordshire SG13 8NL, UK.

A. D. LIQUORISH
Director, Textile Industrial Components Ltd, Tico Works, Hipley Street, Old Woking, Surrey GU22 9LL, UK.

A. I. MEDALIA
Group Leader, Fine Particles Research, Cabot Corporation, Billerica, Mass. 01821, USA.

N. A. MILLER
Technical Officer, Shoe and Allied Trades Research Association, Kettering, Northamptonshire NN16 9JH, UK.

M. D. MOORE
Research Scientist, Avon Rubber Co. Ltd, Melksham, Wiltshire SN12 8AA, UK.

J. C. MORAWSKI
Research Worker, Rhône-Poulenc Industries, Research Centre, Décines, France.

L. MULLINS
Director of Research, The Malaysian Rubber Producers' Research Association, Brickendonbury, Hertfordshire SG13 8NL, UK.

D. E. NEWLAND
Professor in the Department of Engineering, University of Cambridge, Cambridge CB2 1PZ, UK.

D. PETTIT
Head of Adhesion and Solings Research, Shoe and Allied Trades Research Association, Kettering, Northamptonshire NN16 9JH, UK.

E. R. PRAULITIS
Research Worker, Department of Applied Physics, Lanchester Polytechnic, Coventry, Warwickshire CV1 5FB, UK.

A. J. REED
Technical Manager, Metalastik Group, Polymer Engineering Division, Dunlop Ltd, Leicester, Leicestershire LE5 5LY, UK.

R. A. SMITH
Technologist, Avon Rubber Co. Ltd, Melksham, Wiltshire SN12 8AA, UK.

E. SOUTHERN
Senior Lecturer, National College of Rubber Technology, The Polytechnic of North London, London N7 8DB, UK.

A. G. THOMAS
Head of Applied Physics Group, The Malaysian Rubber Producers' Research Association, Brickendonbury, Hertfordshire SG13 8NL, UK.

J. THORPE
Design Engineer, Metalastik Group, Polymer Engineering Division, Dunlop Ltd, Leicester, Leicestershire LE5 5LY, UK.

D. M. TURNER
Director of New Projects, Avon Rubber Co. Ltd, Melksham, Wiltshire SN12 8AA, UK.

I. V. F. VINEY
Senior Lecturer, Department of Applied Physics, Lanchester Polytechnic, Coventry, Warwickshire CV1 5FB, UK.

A. VOET

Research Worker, University of the Upper Rhine, 68200 Mulhouse, France.

W. C. WAKE

Consultant and Visiting Professor, Department of Chemistry, City University, London EC1V 4PB, UK.

R. E. WETTON

Senior Lecturer in Polymer Science, Department of Chemistry, Loughborough University of Technology, Leicestershire LE11 3TU, UK.

R. E. WHITTAKER

Manager, Product Research and Technical Services, Shoe and Allied Trades Research Association, Kettering, Northamptonshire NN16 9JH, UK.

D. C. WRIGHT

Head of Engineering, Rubber and Plastics Research Association, Shawbury, Shropshire SY4 4NR, UK.

CONTENTS

Chapter 1

MECHANICAL BEHAVIOUR OF POLYMERS—ACCOMPLISHMENTS AND PROBLEMS

L. MULLINS

1.1 INTRODUCTION

The use of rubber as an engineering material depends on (i) an understanding of its mechanical behaviour under rapidly changing stresses, and (ii) the development of proper methods of measuring this behaviour, and its dependence on such factors as temperature, frequency, and rate of deformation. It was in these areas that Dr A. R. Payne made his most important accomplishments.

1.2 DYNAMIC BEHAVIOUR OF FILLED RUBBERS

Payne's observation that the dynamic modulus of filled rubber vulcanisates (i) had unexpectedly high values at very small strains (smaller than those normally used), (ii) decreased with increasing strain amplitude up to strains of about 0·1 per cent, (iii) approached an asymptotic value at higher strains, and (iv) in subsequent measurements made at smaller strains was substantially constant and independent of strain amplitude and had a value determined by the highest previous strain, brought a new approach to methods of characterising the stiffening effect of fillers and a new concept to discussions on the reinforcement of rubber by fillers [1].

Typical curves of the change of the in-phase dynamic shear modulus (G') and the out-of-phase dynamic shear modulus (G'') with strain amplitude carbon-black-filled rubber vulcanisates are shown in Figs. 1.1 and 1.2.

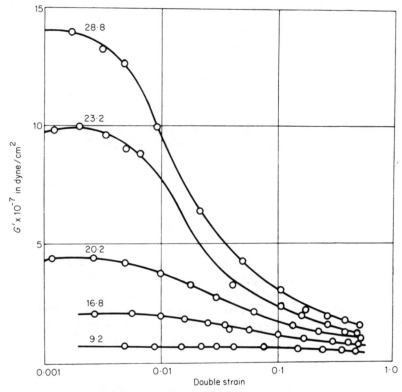

FIG. 1.1. Change of dynamic in-phase shear modulus G' with shear amplitude for a range of butyl–HAF black vulcanisates [18].

The in-phase modulus can be characterised by G_0' and G_∞' for zero and infinite (large) strains, respectively. In rubbers without fillers the change in G' with strain amplitude was found to be very small. Thus the decrease in G' with strain amplitude was attributed to a breakdown in filler structure and the difference $(G' - G_\infty')$ is the contribution of the filler structure to the in-phase shear modulus. On resting after deformation the value of (G') recovers towards its initial value, reflecting reformation of the filler structure.

The value of $(G_0' - G_\infty')$ represents the maximum contribution of the filler structure to the in-phase shear modulus. Its magnitude varies markedly with (i) the type of filler—being highest for those carbon blacks which are commonly regarded as structure blacks, (ii) the

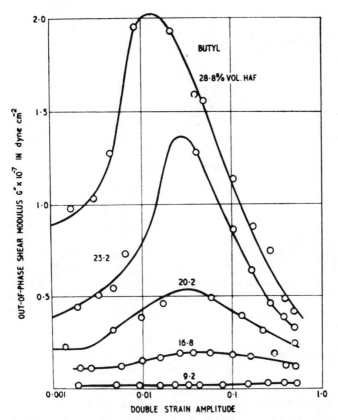

FIG. 1.2. Change in dynamic out-of-phase shear modulus G'' with shear amplitude for a range of butyl–HAF black vulcanisates [18].

concentration of filler, and (iii) the method of mixing the filler with rubber—particularly the incorporation of surface-active chemicals.

However, Payne showed that if the data were reduced by using the ratio $(G' - G'_\infty)/(G'_0 - G'_\infty)$ then its change with strain amplitude was closely similar for different types and different concentrations of fillers and for different rubbers [2]. He claimed that even closer correspondence of behaviour was obtained if, instead of strain or stress as the variable, the product of stress and strain—called 'strainwork' by Payne—was used. Figure 1.3 shows a mastercurve obtained in this way.

This rather surprising observation established that the fraction of

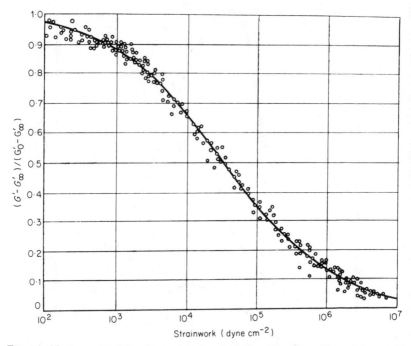

FIG. 1.3. Mastercurve of the change in $(G' - G'_\infty)/(G'_0 - G'_\infty)$ with strainwork for a wide range of filled vulcanisates [18].

the potential of the filler structure to enhance the modulus remaining after the rubber had been strained was determined only by the work done on the system, and that differences in the nature of the filler were of little importance. It also confirmed that the rubber itself did not play any part in this structure.

To describe the observed dependence of $(G' - G'_\infty)$ Payne suggested that the filler structure was a three-dimensional lattice or network of chains or agglomerates of filler particles held together by Van der Waals' bonds of attraction. Although the bonds differed in number—as the structure was more or less well developed depending on the degree of aggregation and dispersion of the filler and thus depending on the concentration and method of dispersion of the filler and type of rubber—they did not differ greatly in their nature. He demonstrated that this filler structure was still evident if the rubber was replaced by a liquid [3], and Fig. 1.4 shows the dependence of the in-phase modulus on strain for a filler–liquid system. This is closely

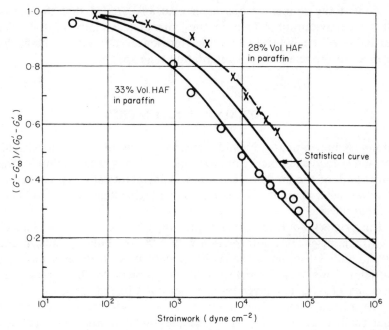

FIG. 1.4. $(G' - G'_\infty)/(G'_0 - G'_\infty)$ versus strainwork for HAF black/paraffin oil mixtures [18].
See Fig. 39 in ref. 18.

similar to that of a filler–rubber vulcanisate (Fig. 1.1). Attempts to describe the physical nature of this structure more precisely have not resulted in any notable advance in understanding. Nevertheless, the technique of describing filler structure in terms of the dependence of (G') on strain amplitude has been of value in characterising different fillers, particularly in assessing their ease and goodness of dispersion and the influence of processing aids on dispersion. It is also of direct importance in the selection of rubber compounds for engineering applications. In such applications changes in mechanical properties are normally unwelcome and reduction of the values of $(G' - G'_\infty)$ and (G'') to a minimum are targets to be aimed at, so as to provide rubber vulcanisates with the best combination of constancy in modulus with minimum creep, set and heat build-up.

The behaviour of the out-of-phase shear modulus (G'') is more complex. It cannot readily be split into two separate components related to filler and rubber respectively, and its value reflects the energy dissipated both in breakdown of filler structure and in deformation of the rubber. It has been plausibly argued that the visco-elastic behaviour of rubber in the immediate vicinity of the filler particles will be influenced by their presence and will be shown by changes in the relaxation or retardation time spectra.

1.3 STRAIN AMPLIFICATION

Although the in-phase dynamic shear modulus of filled rubber vulcanisates decreases markedly with increase in strain amplitude, the lower limit of the modulus G'_∞ which is asymptotically approached at strains greater than about 1 is still much higher than that of rubbers without fillers.

The primary sources of the higher modulus of filled vulcanisates are (i) the absence of deformation within the rigid filler particles, and (ii) the immobilisation of the rubber matrix at the surface of the filler particles. This is a formally identical problem to the increase in viscosity of a liquid caused by a suspension of solid particles.

The modulus thus depends on the concentration and shape of the filler particles as described, for example, by Guth–Einstein equations for spherical (eqn 1.1) and asymmetric (eqn 1.2) particles, respectively

$$E = E_0(1 + 2 \cdot 5c + 14 \cdot 1c^2) \tag{1.1}$$

and

$$E = E_0(1 + 0 \cdot 67fc + 1 \cdot 62f^2c^2) \tag{1.2}$$

where c is the volume concentration of the filler, f is a factor describing the shape of the asymmetric particles as expressed by the ratio of their length to their diameter, E_0 is Young's modulus of the unfilled rubber and E is Young's modulus of the filled rubber.

Good agreement with the predicted dependence on concentration has been obtained with rubbers containing thermal carbon black, which consists essentially of spherical particles. With more reinforcing carbon blacks, fair agreement with eqn (1.2) was obtained using shape factors f of about 6 (Fig. 1.5) [4].

The simplest interpretation of these observations is that the ratio of stress to strain is increased by the presence of filler by a factor X, and, for small deformations and for spherical particles

$$X = E/E_0 = 1 + 2 \cdot 5c + 14 \cdot 1c^2 \tag{1.3}$$

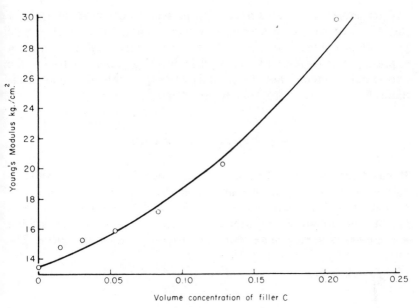

FIG. 1.5. Young's modulus of natural rubber–MT black vulcanisate as a function of concentration [4].

For small deformation, E is the ratio of the stress to the strain, but for larger deformations, the stress and the strain are no longer directly proportional and it is necessary to consider whether it is the stress, or the strain, or a function of both the stress and the strain which is influenced by the presence of the filler. Mullins and Tobin argued that it was more reasonable to consider that the local strains are on average X times greater than the overall strain [4]. They called the ratio X the strain-amplification factor and used it to analyse stress–strain data obtained in simple extension up to large deformations on rubbers containing different volume concentrations of a variety of carbon blacks.

The analysis of their results was consistent with the concept that, to a first approximation, the stress–strain properties of filled rubber vulcanisates could be described in terms of those of the corresponding unfilled vulcanisate if the effective average deformation in the rubber matrix was taken as equivalent to X times the measured overall deformation, where X was determined experimentally as the ratio of the Young's moduli.

Alternative expressions have been derived to describe the ratio of the effective to the measured strain in a filled rubber vulcanisate [5–7]. It is not easy to resolve the relative merits of the different expressions put forward, such differences as exist between the expressions having not been isolated experimentally, and leads naturally to the next problem, stress softening.

1.4 STRESS SOFTENING

It has been observed for many years that deformation results in softening of rubber and that the initial stress–strain curve on a new sample of rubber is unique (Fig. 1.6). Most of the softening occurs in the first deformation and after a few deformation cycles the rubber approaches a steady state with a constant stress–strain curve. The

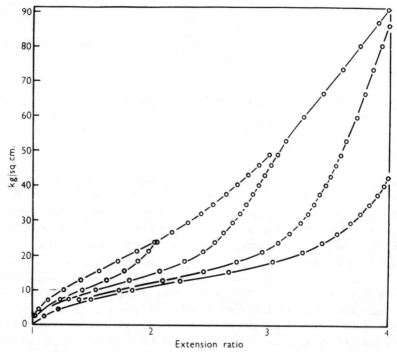

FIG. 1.6. Effect of previous stretching on load–extension curves of natural rubber–MPC black vulcanisate [19].

softening is usually only present at deformations smaller than the previous maximum. Softening in this way occurs in vulcanisates with or without fillers although the effect appears to be much more pronounced in vulcanisates containing high proportions of reinforcing fillers [8].

This phenomenon has been termed the Mullins effect and has been the subject of numerous investigations. Various qualitative theories have been advanced to describe the effect and as it appears to be more pronounced in filled vulcanisates, discussion has mostly concentrated on a breakdown of filler structure. Thus the softening has been attributed to breakdown or slippages of linkages between filler and rubber, breakdown of filler–filler aggregates, and breakdown of network chains. Although each mechanism will lead to softening, there is an absence of direct evidence to identify their individual contributions to the softening process.

New light was thrown on the source of the softening of both filled and unfilled vulcanisates resulting from previous stretching by the observation of Harwood, Mullins and Payne [9] that when allowance was made for the strain amplification resulting from the presence of filler, the degree of softening was similar in magnitude for both filled and unfilled rubbers. Figure 1.7 shows typical stress–strain curves obtained by them during repeated extension and retraction of natural rubber vulcanisates both with and without filler. The vulcanisates were deformed at constant rate to the same maximum stress and thus to similar maximum effective strain. The close correspondence between the softening in the filled and unfilled vulcanisates when compared at the same stress led them to use the expression 'stress softening' to describe this phenomenon and they concluded that the softening process was primarily due to the rubber phase itself and that the contribution of the filler to the softening was relatively small (Fig. 1.8). Subsequent studies by Harwood and Payne confirmed these conclusions [10, 11] (Fig. 1.9).

These observations led to the conclusion that stress softening was primarily a result of rearrangement of the configuration of the molecular networks. It involved displacement of network junctions and entanglements from their initial random positions—owing to localised non-affine deformation as shorter network chains became fully extended—and resulted in less stress being required to produce deformations smaller than the previous deformation.

In filled vulcanisates the majority of the softening was due to the

L. Mullins

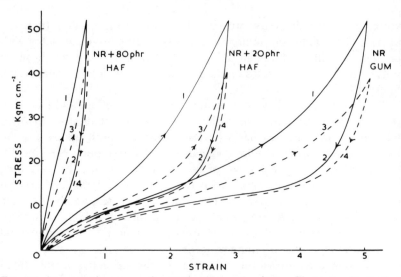

FIG. 1.7. Stress–strain curves of natural rubber vulcanisates. First extension 1; first retraction 2; second extension 3; second retraction 4 [9].

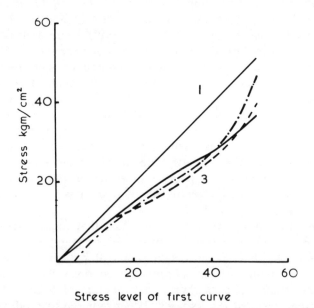

FIG. 1.8. Stress–strain curves in Fig. 1.7 obtained in first extension (1) and second extension (3) re-plotted as stress against stress in first extension (—, NR; – – –, NR + 20 phr HAF black; – · –, NR + 80 phr HAF black).

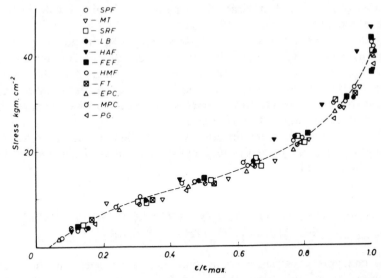

FIG. 1.9. Stress–strain curves obtained in second extension of natural rubber unfilled and filled vulcanisates with strain plotted as a fraction of the maximum strain [9].

same mechanism. Alternative mechanisms involving breakdown of filler–filler and filler–rubber linkages were considered to be relatively unimportant and generally only responsible for a minor part of the softening. However, in these vulcanisates, filler particles and aggregates would also be displaced and oriented in the network during deformation and explain the slower and less complete recovery of the stress softening of filled vulvanisates compared with the rapid and almost complete recovery in unfilled vulcanisates. In some vulcanisates with weak crosslinks there is evidence of more permanent softening owing to breaking of network junctions.

A relatively simple picture now emerges. Stress softening resulting from previous deformation is a frequent and widespread phenomenon. It is observed in polymeric materials with and without fillers and reflects configurational changes within the fine structure of the material which permits subsequent deformation to take place more readily.

In filled vulcanisates, pronounced softening may also take place at very small deformations owing to the breakdown of weak aggregates of filler particles. This breakdown is essentially complete at small strains (*ca.* 0·1) and at larger deformations softening owing to this cause is relatively small.

Although the main sources of stress softening in polymers now appear to be correctly identified, the problem of characterisation of stress–strain behaviour remains complex. It appears that the initial stress–strain curve is unique and that in subsequent deformation the stress related to each strain reflects elements of the memory of previous deformations to higher strains.

The use of a strain-amplification factor to describe the increase in modulus resulting from the incorporation of fillers is an important unifying concept.

1.5 APPLICATION OF STRAIN AMPLIFICATION AND STRESS SOFTENING TO OTHER PROPERTIES

The concepts of strain amplification and stress softening in filled rubbers have also been shown to have important relevance to measurements of other properties such as breaking elongation, set and creep. Figure 1.10(a) (From Harwood, Payne and Whittaker [12]) shows the energy at break of unfilled and carbon-black-filled vul-

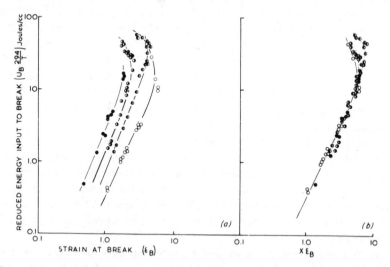

Fig. 1.10. Variation of energy at break with strain at break for SBR vulcanisates containing 0, ○; 30, ◑; 60, ◑; and 80, ● phr HAF black [12].

canisates plotted against the actual strain at break. Figure 1.10(b) shows the same data re-plotted against the effective strain at break as determined by the use of the strain-amplification factor, X. All the data are now described by a single curve, thus confirming the value of the strain-amplification factor to account for the behaviour of the rubber matrix in filled vulcanisates.

Results obtained by Harwood and Payne [13] on the permanent set of both unfilled and carbon-black-filled rubbers are shown in Fig. 1.11. Here again use of the strain-amplification factor reconciles data obtained on both unfilled and filled rubbers.

Recent studies by Derham and Thomas [14] of the creep under repeated stressing of both unfilled and filled rubbers have shown a more complex situation. Figure 1.12 shows typical data obtained during both the cyclic and static loading of a carbon-black-filled natural rubber vulcanisate with a load producing an initial extension of 100 per cent. The creep under repeatedly applied tensile loads is very much larger than would be expected from observation of the creep under the same load applied continuously. This is not a simple stress-softening effect as it is well known that the static creep of many rubbers, especially those containing fillers, decreases if the rubber is pre-stressed.

The creep behaviour of a natural rubber vulcanisate without fillers is more closely in accord with expectations and at lower strains (less than about 100 per cent) the cyclic creep rate is less than the static creep rate (Fig. 1.13) and the rubber approximately obeys the Boltzmann superposition principle.

If account is taken of the strain-amplification effect present in filled rubbers and the creep rates of both the filled and unfilled rubbers are compared at equal stress, then the differences between their creep rates are much reduced, but a substantial difference still remains (Fig. 1.14), and an additional reason must be sought for these differences in behaviour.

Derham and Thomas suggested that the phenomenon was associated with crystallisation of natural rubber under strain. They used as an analogy the crack-growth behaviour of natural rubber, where repeated stressing produces crack growth, but a static load does not. In the case of the latter, strain-induced crystallisation at the tip of the crack prevents bond rupture continuing to its full extent, but after relaxation more bond rupture takes place in a subsequent extension before crystallisation again occurs. They claimed that the observation

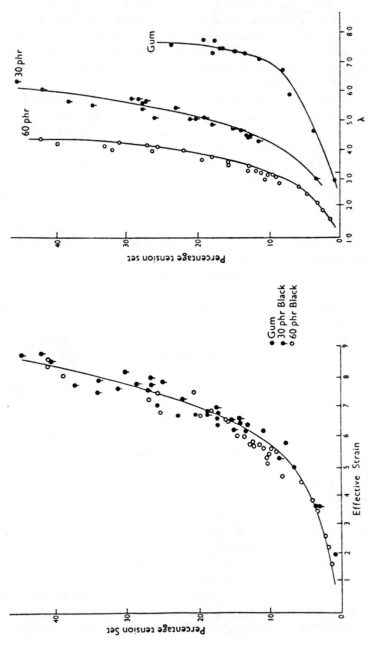

FIG. 1.11. Change of permanent tension set with extension ratio for natural rubber vulcanisates containing 0, 30 and 60 phr HAF or ISAF black [13].

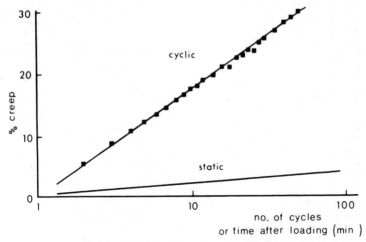

FIG. 1.12. Cyclic and static creep of natural rubber vulcanisate containing 40 phr carbon black [14].

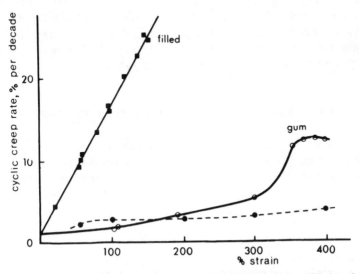

FIG. 1.13. Dependence of cyclic creep rate on initial strain for unfilled and filled vulcanisates. Also shown is the static creep rate of unfilled vulcanisate [14].

L. Mullins

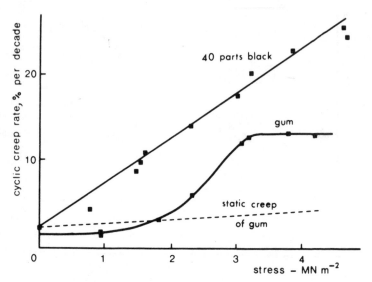

FIG. 1.14. Cyclic and static creep rates as functions of applied stress [14]. 1 MN · m^{-2} = 1 MPa.

that the cyclic creep was much reduced if measurements were made under non-relaxing conditions strengthened the analogy. This interpretation implies that bond rupture or slippage in regions of high stress concentrations at or near the rubber–filler interface are responsible for the enhanced cyclic creep of filled rubbers and that this is inhibited by strain-induced crystallisation.

1.6   HYSTERESIS AND STRENGTH OF RUBBERS

For many years—certainly since the development in the mid 1950s of the tearing energy approach to describe rupture—it has been recognised that hysteresial processes, which occur at the very high strains associated with the propagation of rupture, make an important contribution to the strength of materials, as evidenced by the observation that strength depends upon rate and temperature in a manner which parallels visco-elastic behaviour [15].

The observation by Grosch, Harwood and Payne [16] that there was a simple relation between the energy required to break rubber (U_B) in a tensile test and the hysteresial energy loss at break (H_B)

suggested that the relationship was worthy of more detailed consideration. The relation

$$U_B = KH_B^{2/3} \tag{1.4}$$

applied to a wide range of amorphous rubber in which the hysteresis was varied by changes in temperature and rate and by swelling rubber to different extents (Fig. 1.15).

Values of the constant K were different for different polymers. With filled rubbers, although the 2/3 relationship still held, the results for different fillers provided a series of parallel lines but Harwood and Payne found by the introduction of the strain-amplification factor X,

$$U_B = K(H_{B/X}^{2/3}) \tag{1.5}$$

that it was possible to reduce all the data to a single line [17].

Although attempts have been made to explain this empirical relationship they have not achieved much success. It is difficult to see its physical significance. The explanation must be sought in the effect of hysteresis on the stresses encountered by the tip of the growing crack

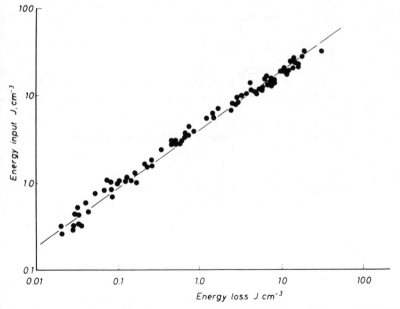

FIG. 1.15. Relation between breaking energy and hysteresial energy loss at break [20].

and the energy losses in the rubber in the neighbourhood. Progress in
this direction will probably make the use of a relationship involving
the use of bulk properties of this type inappropriate.

*Bob Payne's many contributions provide a lasting record to his
success as a scientist, but an even stronger memory, to all of us who
knew him, is that of Bob Payne as a person. He was a man of
tremendous energy, and, although he did not enjoy the best of health,
he drove himself apparently untiringly. He was thoughtful and kindly
to others and was always ready to be interested in and help in their
problems, and it is for this that he is missed.*

REFERENCES

1. A. R. PAYNE. *Rubb. Plast. Age*, 1961, **42**, 963; *J. Appl. Poly. Sci.*, 1962, **6**, 57; *J. Appl. Poly. Sci.*, 1963, **7**, 873.
2. A. R. PAYNE. *J. Appl. Poly. Sci.*, 1965, **8**, 2661.
3. A. R. PAYNE. *J. Coll. Sci.*, 1964, **19**, 744.
4. L. MULLINS and N. TOBIN. *J. Appl. Poly. Sci.*, 1965, **9**, 2293.
5. A. M. BUECHE. *J. Poly. Sci.*, 1957, **23**, 139.
6. F. BUECHE. *J. Appl. Poly. Sci.*, 1961, **5**, 271.
7. Y. SATO and J. FURUKAWA. *Rubb. Chem. Tech.*, 1962, **35**, 857.
8. L. MULLINS. *J. Rubb. Res.*, 1947, **16**, 275; *J. Phys. Coll. Chem.*, 1950, **54**, 239.
9. J. A. C. HARWOOD, L. MULLINS and A. R. PAYNE. *J. Appl. Poly. Sci.*, 1965, **9**, 3011.
10. J. A. C. HARWOOD and A. R. PAYNE. *J. Appl. Poly. Sci.*, 1966, **10**, 315.
11. J. A. C. HARWOOD and A. R. PAYNE. *J. Appl. Poly. Sci.*, 1966, **10**, 1203.
12. J. A. C. HARWOOD, A. R. PAYNE and R. E. WHITTAKER. *J. Appl. Poly. Sci.*, 1970, **14**, 2183.
13. J. A. C. HARWOOD and A. R. PAYNE. *Trans. I.R.I.*, 1966, **42**, 14.
14. C. J. DERHAM and A. G. THOMAS. *Rubb. Chem. Tech.*, 1977, **50**, 397.
15. L. MULLINS. *Trans. I.R.I.*, 1959, **35**, 213; K. A. GROSCH and L. MULLINS. *Rev. Gen. Caout.*, 1962, **39**, 1781,
16. K. A. GROSCH, J. A. C. HARWOOD and A. R. PAYNE. In: *Inst. Phys. and Phys. Soc. Conf.*, 1966, series No. 1, p. 144.
17. J. A. C. HARWOOD and A. R. PAYNE. *J. Appl. Poly. Sci.*, 1968, **12**, 889.
18. A. R. PAYNE. (1965). *Complex Systems*, Ch. 7. Amsterdam: Elsevier.
19. L. MULLINS and J. TOBIN. *Trans. I.R.I.*, 1956, **33**, 1.
20. J. A. C. HARWOOD and A. R. PAYNE. *Calver's Group Français de Rheologie*, 1966, 207.

Chapter 2

DESIGN OF ELASTOMERS FOR DAMPING APPLICATIONS

R. E. WETTON

2.1. INTRODUCTION

Elastomeric materials are used in dynamic engineering in two completely different ways. In one case elastomers with high damping are required for application to structures to damp out residual resonances still remaining after optimising mass distributions. Common examples of this type are found in transport engineering with largely cheap bituminous felts used for automobile panel damping but with more weight efficient materials used in special situations, such as aircraft structures. The vital need for adequate suppression of resonant modes has long been recognised in the aircraft industry, to avoid catastrophic noise or stress induced vibration failure.

The other use of elastomers in dynamic engineering is in the field of elastic mountings and suspensions in which the frequency characteristics of the rubber are a vital part of the whole design. In very few cases have the frequency and damping characteristics achievable from polymers been used scientifically in this context. The engineer thinks in terms of low-hysteresis rubbers for springs and although tricks are played with linearity and anisotropic stiffness, the frequency characteristics of damping rubbers are rarely exploited.

As will be seen in this paper, the dynamic frequency characteristics of polymer systems could be an excellent asset to a given engineering component, but suffer from the serious disadvantage that these characteristics vary with temperature. The main theme of this paper is thus to show that recent developments and discoveries can produce damping elastomers in which the temperature dependence is minimal or at least can be reduced to acceptable levels.

19

2.2 GENERAL CONSIDERATIONS FOR THE USE
OF DAMPING ELASTOMERS

2.2.1 Vibration damping applications

It is not possible in a short article to consider even a significant number of the many types of geometrical structures requiring damping. The principles are therefore discussed in terms of the damping of plates, which provide the most common application of added damping. The plate could be equated with aircraft fuselage panels or air ducting in buildings. A plate can vibrate with a large number of different modes but generally the mean square of the amplitude of vibration $(\overline{\gamma^2})$ will be given by a relation of the type

$$\overline{\gamma^2} = \frac{\pi S(\omega)}{M \omega_r K''} \tag{2.1}$$

where $S(\omega)$ is the spectral power density of the excitation source, M the vibrating mass, ω_r the resonant frequency and K'' the out-of-phase force constant of the system, *i.e.* if (stress amplitude/unit displacement) $= K^*$, then

$$K'' = K^* \sin \delta \tag{2.2}$$

where δ is the loss angle. This expression states a generally true rule that for a flat excitation power spectrum, the lower the resonant frequency, the larger is the amplitude of the resonance excited. Thus, the lower frequencies may need to be damped to avoid structural damage. The higher-frequency vibrations (50 Hz–20 kHz) are however in the audio range and must therefore be damped to reduce subjective noise nuisance. Table 2.1 shows two common ways of achieving the damping of panels by unconstrained damping layers [1] and by damping layers constrained by a metal foil on the surface. The relationship given for the constrained [2] case is only approximate and assumes that the constraining foil is thin compared with the plate being damped and also is of the same modulus.

The relations given in Table 2.1 show that the requirement for maximising damping with unconstrained elastomeric layers is primarily for the damping layers to have a high loss component of Young's E'' modulus. This arises physically because the layer is stretched and compressed as a plate bends. The situation in the constrained layer however requires not only a high tan δ for the damping layer but also a *low* storage component of the rigidity

TABLE 2.1
Modal damping of plates

1. *Unconstrained layer:* requires damping layer loss modulus as high as possible		$\tan \delta = \dfrac{3E''_d t_d}{E'_m t_m}$ where E''_d = loss modulus of damping layer (Young's storage modulus); E'_m = modulus of metal plate; t_d, t_m = thicknesses of damping layer and plate, respectively.
2. *Constrained layer:* requires $\tan \delta_d$ to be high but the modulus of the damping layer to be low to maximise f(g)		$\tan \delta = \tan \delta_d \times \dfrac{12a^2 b f(g)}{1 + (12a^2 + 2)b \, f(g)}$ where $\tan \delta_d$ = damping factor for damping layer material; a is the ratio of damping layer to plate thicknesses; b is the ratio of foil-to-plate stretching stiffness and g is a shear parameter

modulus (G'). This then allows an optimum amount of shear to occu
in the constrained layer.

2.2.2 Elastomeric suspensions with defined damping and frequency characteristics

Rubber components are frequently used as anti-vibration mounting
for machinery and sometimes for buildings, with earthquake protec
tion one of the goals of current research. The aim of these mounting
is usually, although not always, to provide isolation rather tha
damping. In special cases a frequency-dependent stiffness and part
cular damping characteristics are required. A selection of application
of elastic suspensions with the required dynamic properties is show
in Table 2.2. The interesting fact emerges that for the majority c
dynamic applications, a rubber spring is required to be soft at lo
frequencies but with the stiffness (proportional to the dynami
storage modulus) increasing with frequency. This rule arises becaus
large amplitude motion is normally only required at low frequencie:
Examples are stylus tracking at low frequencies and the large-scal
motions involved in vehicle cornering, which are again low-frequenc
motions. A reasonable level of damping is desirable at these frequer
cies to suppress low-frequency resonant modes. At high frequencie
the spring units in general need to be stiffer to push norm;
resonances above the possible range of excitation frequencies and t

TABLE 2.2
Elastic mountings—damping requirements

Application	Dynamic property
Vehicle suspension components (*e.g.* yaw suspension in rail vehicles)	Stiffness increasing with frequency; damping increasing with frequency
Overhead electric pantograph mounting	Stiffness increasing with frequency; damping increasing with frequency
Gramophone stylus mounting	Stiffness increasing with frequency; damping increasing with frequency
Anti-vibration bearing	Soft in shear, stiff in compression damping medium/low, for isolation

minimise 'hunting modes' in the case of the yaw suspension unit for a wheel-set in rail vehicles. High damping is desirable to suppress any high-frequency resonances which do occur.

2.3 DYNAMIC MECHANICAL CHARACTERISTICS OF ELASTOMERS—THE PROBLEM OF TEMPERATURE DEPENDENCE

The major damping in polymers occurs in a temperature region not far above the glass transition (T_g) when the rate of configurational rearrangements of the chains is of the same order as the impressed frequency. Thus, polymers with a low T_g value in general have rapid molecular motion at room temperature with low damping at normal frequencies. The extrapolated frequencies of maximum damping of a range of elastomers at room temperature are given in Table 2.3 together with their T_g values. The first two elastomers (natural rubber and SBR) are essentially non-damping (low hysteresis) at normal frequencies, but the other materials are elastomers with potentially useful damping characteristics. In particular polyurethanes can be synthesised with a wide range of T_g values by varying the chemical nature of the 'soft' segment.

In Fig. 2.1 the dynamic properties of poly(isobutylene) are shown in the frequency plane at a number of different temperatures. The data were obtained on the PLL prototype dynamic mechanical thermal

TABLE 2.3
Loss peak locations for selected elastomers

Polymer	$T_g(°C)$	Frequency of damping maximum at room temperature
Natural rubber	−73	~10 GHz
SBR (23·5% styrene)	−63	~100 MHz
Poly(isobutylene)	−68	~10 kHz
Polyurethanes (polyester based)	−70–+70	d.c.–1 GHz
Polynorbornene	+40	
Polynorbornene (plasticised)	−40–+40	d.c.–10 kHz

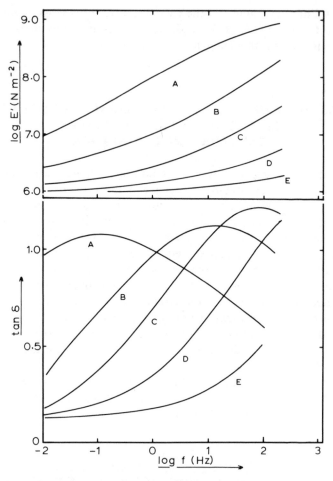

FIG. 2.1. Dynamic Young's modulus (E') and damping (tan δ) as a function of frequency for unplasticised, lightly crosslinked poly(isobutylene). The temperatures (°C) of measurement were: A, $-51\cdot5$; B, $-31\cdot0$; C, $-16\cdot5$; D, $+1\cdot0$; E, $+24\cdot0$.

analyser (DMTA) [3]. This particular elastomer in the unfilled state may be compounded to produce the required increasing storage modulus with frequency, coupled with damping characteristics which make it suitable for gramophone stylus mountings. However Fig. 2.1 immediately reveals the major problem, aside from environmental effects which are not considered in this paper, in the application of elastomers with useful dynamic characteristics. The high damping and

high rate of change of storage modulus with frequency are coupled to
a high temperature sensitivity. Thus if characteristics are optimised at
one temperature, a change of only 10°C can render them far from
ideal. This temperature sensitivity originates directly from the increasing rate of molecular rearrangement with temperature and is a
feature of all mechanical characteristics deriving from the glass–
rubber transition.

2.4 DESIGN OF ELASTOMERS FOR DAMPING APPLICATIONS

2.4.1 Loss peak broadening

If very high levels of damping are not required, say tan $\delta \simeq 0\cdot2$–$0\cdot4$
then a number of compromises can be reached with the temperature
problem. The damping peak can be broadened to an almost indefinite
degree, at lower tan δ levels, so that a shift in the peak with change of
use temperature does not materially affect the damping level.
However, changes in the frequency characteristics of storage modulus cannot be obviated and although the rate of change of ln E' with
frequency (f) may be similar in the approximation of Staverman and
Schwarzl [4]

$$\tan \delta = 2\pi \cdot \frac{d(\ln E')}{d \ln f} \tag{2.3}$$

the absolute values will decrease as temperature increases.

In general, a damping peak can be broadened by blending
components which are inherently incompatible to form a basically
single phase. Under these conditions localised fluctuations in concentration occur on the 10–50-Å scale and generate a range of relaxing
environments [5]. The temperature location of the damping peak
(T_{max}) under these conditions is given approximately by

$$T_{max} = x_1 T_{max}(1) + x_2 T_{max}(2) \tag{2.4}$$

where x_1 and x_2 are the mole fractions of the two components which
have damping peaks at $T_{max}(1)$ and $T_{max}(2)$, respectively.

If incompatibility is too great or if the components are not
sufficiently constrained to co-exist in a single phase, then separate
relaxation peaks are observed. This trend in the degree of molecular
mixing is shown in Fig. 2.2 for a series of polybutadiene-containing
elastomers. A sharp relaxation peak with high damping values is

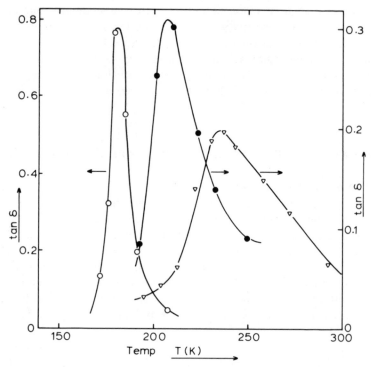

FIG. 2.2. Series of polybutadiene-containing elastomers with damping curves broadening with increasing complexity of the molecular environment. Styrene–butadiene–styrene block copolymer with good phase separation, ○; random styrene–butadiene copolymer, ●; same random copolymer 76 per cent hydrogenated, ▽.

observed for the polybutadiene component of a completely phase-separated block copolymer (styrene–butadiene–styrene, Shell K 1101). Instead of incorporating the butadiene segments in long blocks, the same monomers can be randomly copolymerised into the same chain, thus preventing large-scale phase separation. The random copolymer then has a broadened damping with its location intermediate with the peaks for the parent polymers.

The range of molecular environments can be made even more complex by partially hydrogenating the butadiene [6] to give chains which are essentially random terpolymers of styrene, butadiene and ethylene. The frequency half-width of the damping peak for this material is now six decades as opposed to the original two decades or the original polybutadiene phase. The tan δ magnitudes, as can be seen in Fig. 2.2, are progressively reduced as the curves are broadened.

Hirsh has recently [7] tried to correlate in a practical way the level of damping in a number of elastomers with the rate of change of E' with temperature. This correlation is shown in Fig. 2.3 in terms of the temperature interval ΔT required to produce an increase of 50 per

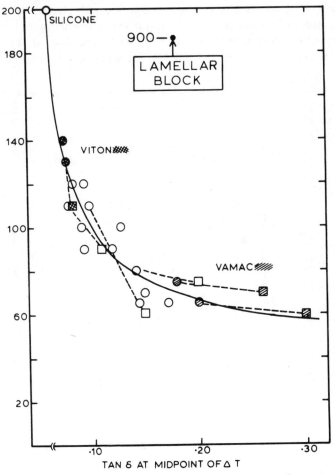

FIG. 2.3. Correlation of tan δ level with temperature dependence of dynamic modulus (E'). ΔT is the temperature interval required for a 50 per cent increase in E' from a point 160°C above T_g (approx.). (After Hirsch [7].)

cent in E' (starting at $T_{max} + 160°C$) versus tan δ at mid-temperature range for this material. Thus a large ΔT indicates a relatively flat response of E' to changing temperature. However, the elastomers with a flat E' versus temperature response show, as expected, low tan δ values. This correlation exists basically as the temperature plane version of the Staverman and Schwarzl relation, knowing that there is a general relationship between frequency and temperature, as embodied in the Williams, Landel and Ferry equation [8].

A material which has below average rate of change of E' with temperature (*i.e.* large ΔT in Fig. 2.3) is the ethylene methylacrylate-acrylic acid terpolymer (VAMAC) [9]. The variations of the moduli and damping of this polymer with temperature are shown in Fig. 2.4,

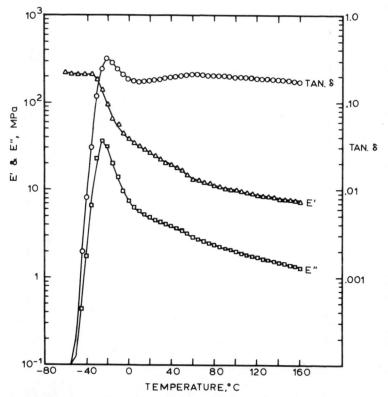

FIG. 2.4. Dynamic data (11 Hz) for plasticised VAMAC compounded in the following parts: VAMAC, 124; FEF black, 50; dioctyl sebacate, 10; Santicizer-409, 10; Armeen 18-D, 0·5; stearic acid, 2; Diak-1, 1·25; diphenylguanidine, 4. (After Hirsch [7].)

where it can be seen that for this optimum formulation the damping is effectively constant, at about 0·2, from 0 to 160°C, but of course the modulus still decreases with temperature increase. The origin of this protracted damping region is a series of overlapping transitions rather than one broadened process. In particular the main glass transition, polyethylene microcrystallite melting and structural breakdown of the ionic clusters [10] may all contribute.

Another system which has a useful range of damping peak locations is a HAF black formulation of polynorbornene, whose structure is shown in Fig. 2.5. This is a relatively new material from Charbonier de France (France) and although a glassy state material in the unplasticised form, it can be plasticised by naphthenic or paraffinic oils to allow reasonably good choice of the damping peak location. Some measurements with this polymer are reported later.

2.4.2 Lamellar block copolymers

An interesting observation has recently been made by Wetton and Taminello [11] that a non-molecular damping can be achieved in block copolymers which have separated into rubbery and glassy lamellae. This relaxation is phenomenologically similar to that known to occur in semi-crystalline polymers with lamellar morphology. Figure 2.6 is an electron photomicrograph showing the degree of regularity which can be achieved. In this case the system was an α-methylstyrene–silicone star block copolymer [12] and the glassy lamellae are approximately 80-Å thick, while the predominant polydimethylsiloxane phase is approximately 153-Å thick. Typically, a domain is seen to extend over dimensions of at least 0·1 μm. The system comprises a multitude of domains of different orientations, but generally with a preference for orientation in the plane of casting or moulding. Using a more exact model than that used by us previously we obtain

$$\tau = \frac{5l^2(h + 3d)^2\eta_R}{6dh^3E_G} \tag{2.5}$$

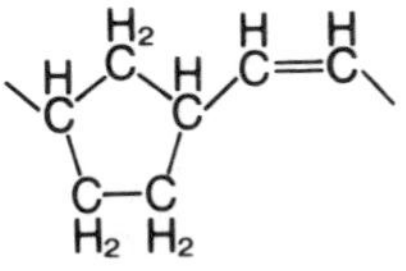

FIG. 2.5. Monomeric unit of polynorbornene. The *trans* isomer will also be present.

FIG. 2.6. Transmission electron photomicrograph of α-methylstyrene–silicone (60 per cent silicone) star block copolymer showing lamellar phase separation with extensive long-range order.

for the relaxation time (τ) of a deformation process where bending of the glassy domains (length, l; thickness h) is resisted by viscous shear in the silicone phase (thickness d) sandwiched between them. Using $\eta_R = G''/\omega$ [13] at $100\,\mathrm{N \cdot s \cdot m^{-2}}$ and $E_G = 3 \times 10^9\,\mathrm{N \cdot m^{-2}}$ for the α-methylstyrene [14], we obtain $\tau = 5 \times 10^{-5}\,\mathrm{s}$, $f_{max} = 3\,\mathrm{kHz}$ for domain length $l = 1000\,\text{Å}$ or $\tau = 5 \times 10^{-4} \times 10^{-3}\,\mathrm{s}$, $f_{max} = 30\,\mathrm{Hz}$ for $l = 1\,\mu\mathrm{m}$. Thus, the damping maximum is predicted to be in the audio range. Experimentally, this is found to be the case with tan δ and E' shown in the frequency plane in Fig. 2.7. These data are shown over $100°$ range about ambient and it is apparent that the damping charac-teristics change only slightly and the modulus essentially not at all. Even closer constancy has been observed in related samples and it is

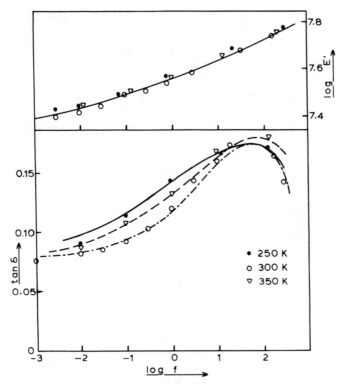

FIG. 2.7. Dynamic mechanical data on α-methylstyrene–silicone star block copolymer with lamellar morphology. The loss process between 10 Hz and 100 Hz is essentially insensitive to temperature.

believed to arise because η_R and E'_G are decreasing at approximately the same rate with temperature. It should be pointed out here that non-lamellar morphology can be induced in these same polymers and in these cases the loss at room temperature is low. The lamellar relaxation would therefore seem to be the best solution so far to the temperature-dependence problem for materials with damping in the 0·1–0·2 tan δ range. This can be seen particularly well on the correlation diagram (Fig. 2.3) in which the point corresponding to the lamellar block copolymer system is at 900 and way off scale on this graph. Materials of this type should prove suitable for some of the dynamic engineering operations outlined in Tables 2.1 and 2.2.

32   R. E. Wetton

2.5   EFFICIENCIES IN VIBRATION DAMPING

In order to test the efficiency of the lamellar block materials as
damping layers a comparison was made with a soft heavily-damped
material, in this case polynorbornene plasticised 50% with a paraffinic
oil. A steel cantilever was coated firstly with a soft polynorbornene
layer, the same layer was then constrained with 0·01-mm aluminium
foil and then after replacement of these with lamellar block

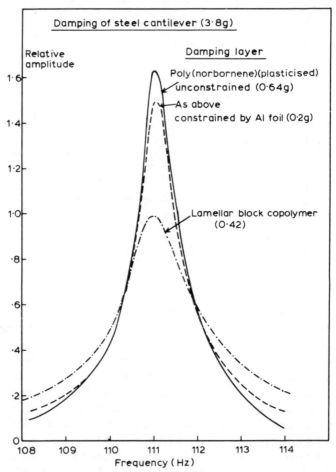

FIG. 2.8. Amplitude of vibration through resonance of steel cantilever damped by various
one-sided treatments as indicated.

copolymer cast to give lamellar morphology. In each case the first fundamental was excited in a vibrating reed apparatus [15] and the amplitudes plotted through the resonance region in each case at room temperature. The results are given in Fig. 2.8. The constrained-layer damping is seen to be somewhat more efficient than that of the unconstrained layer, but the constraining foil is very thin in this case giving a poor t_f/t_m ratio. The damping efficiency, in terms of the effectiveness per gram of damping medium applied, is seen to be far higher in the case of the lamellar block copolymer. This reflects its high loss modulus, which is the requirement for efficient damping with unconstrained geometry.

2.6 CONCLUSIONS

A number of elastomers are now available, in which compromise solutions to the temperature dependence of damping give systems which can be incorporated into dynamic engineering applications. The best results in the low damping range are achieved by lamellar structure block copolymers. These exhibit a non-molecular relaxation which is to all intents and purposes independent of temperature over a 110°C range from −30 to 80°C. They are shown to be highly efficient in damping the vibrations of plate-like structures.

REFERENCES

1. H. OBERST and K. FRANKENFELD. *Acoustics*, 1952, 2, 181.
2. E. M. KERWIN. *J. Acoust. Soc. Am.*, 1959, 31, 952.
3. *Dynamic Mechanical Thermal Analyser (DMTA)*. Polymer Laboratories Ltd, Church Stretton, Shropshire, UK.
4. A. J. STAVERMAN and F. SCHWARZL. (1956). *Die Physik der Hochpolymeren* (Ed. H. A. STUART), Vol. IV, Ch. 1. Berlin: Springer-Verlag.
5. R. E. WETTON, W. J. MACKNIGHT, J. R. FRIED and F. E. KARASZ. *Macromolecules*, 1978, 11, 158.
6. E. W. DUCK, J. R. HAWKINS and J. M. LOCKE. *I.R.I. Journal*, 1972, 6(1).
7. A. E. HIRSCH. (1977). In: *International Rubber Conference*, Vol. II, Brighton, UK, 16–20 May 1977.
8. J. D. FERRY. (1970). *Viscoelastic Properties of Polymers*, 2nd edn, p. 314. New York: John Wiley.
9. 'VAMAC'—trade name of DuPont de Nemours Co., Wilmington, Delaware, USA.

Chapter 3

MECHANICAL PROPERTIES OF RUBBERS RELEVANT TO THE ENGINEERING OF THEIR PROCESSES

D. M. TURNER, M. D. MOORE and R. A. SMITH

3.1 INTRODUCTION

This paper is concerned with exploring the behaviour of unvulcanised filled rubber in situations which arise in the normal run of processes in a factory and trying to provide a consistent explanation so that various types of behaviour and results of different tests can be interrelated.

3.2 A RHEOLOGICAL MODEL

The TMS (Turner, Moore and Smith) model shown in Fig. 3.1 has been developed so that the equations derived from it are capable of predicting this behaviour of unvulcanised rubber in a variety of processing situations. The equations for the basic operations are included in Appendix 3.1. These equations can be readily evaluated by numerical methods.

The model is basically two Maxwell networks in parallel. The need for dual networks has been recognised by Tobolsky [1] and Gent [2] to permit an interaction between one network and the other to give time-dependent recovery effects. The stiffness of the D spring is always several times that of the E spring. The viscous constants K and J usually are within a factor of 4 of each other.

The strength of bonds and interactions which maintain integrity of a piece of unvulcanised rubber are limited and may be exceeded by

35

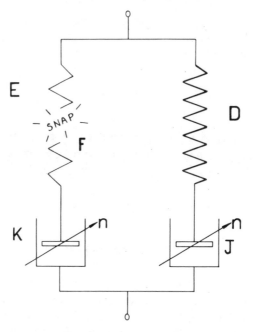

FIG. 3.1. The TMS model.

forces imposed by processing machinery, resulting in material fracture. E and K appear to be associated with the weaker links, *e.g.*, those caused by carbon-black gel and entanglements. When stress F is exceeded in the E/K network, fracture results.

In 1921 Nutting [3] postulated that a power law was appropriate for describing stress relaxation in a large number of polymeric systems. The value of this approach was confirmed by Cotten and Boonstra [4] in 1965. A spectrum of relaxation times can be arranged to give a similar response, but with its large number of constants it is cumbersome to use. Dove, Turner and Martin [5] showed that a power law damper in a Maxwell model leads to power law stress relaxation. Power law damping is incorporated in both networks of the TMS model.

All types of deformations used in rubber processing, be they short, prolonged, shearing or elongating, give rise to forces depending on both elastic and viscous effects.

3.3   APPLICATIONS OF THE MODEL TO RHEOLOGICAL TESTS

3.3.1   Elongation testing

Funt [6] has been an advocate of elongation testing, measuring the force required to stretch a sample of unvulcanised rubber at a constant rate. In Fig. 3.2 the results of such a test are plotted as true stress against elongation ratio. This illustrates very directly the action of the TMS model. The test conditions and parameters used in calculations of model response are given in Appendix 3.2 for this and subsequent figures.

Initially, the response is elastic, hence the rate of increase in stress is given by $D + E$. The contribution of the $D:J$ network is limited by yielding of the J dashpot. In this particular test the extension rate was constant; the elongation rate of the sample dropped as it became extended and hence the stress in the $D:J$ network dropped. At higher

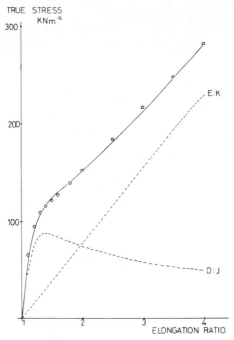

FIG. 3.2. $\bigcirc$, Experimental point; – – – –, predictions from the TMS model of the contributions from each network; ————, the contributions from the two networks combine to fit the experimental points. (See Appendix 3.2.)

elongation ratios any increase in stress will be due to stretching the *E* spring. Eventually the sample fails; sometimes it does this by yielding in an orderly manner, in which case an estimate can be made of *K*, or it may fail in a brittle manner which enables *F* to be determined.

3.3.2 Creep testing
A related test was carried out by Freakley and Wan Idris [7] whereby they applied a constant stress to an extruded cord of rubber and measured the strain. In Fig. 3.3 the results of creep tests are shown on three mixes of the same NR black formulation but having had different quantities of energy input during their mixing. An increase in the energy by a factor of 2·7 causes *E* and *D* to reduce to two-thirds of their original values. Elongation tests have also revealed systematic changes in the two elastic parameters with mixing time which are more pronounced than those in the viscous parameters.

3.3.3 Capillary rheometer testing
Equations can be derived from the model to predict the shear stress required for flow at a given shear rate at the capillary wall. A typical

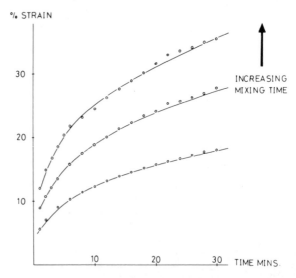

FIG. 3.3. ○, Constant stress creep result for a black filled OENR formulation. The three sets of points differ only in energy input during mixing, the uppermost line having had the most mixing. ———, Predictions from the TMS model with a suitable choice of parameters. (See Appendix 3.2.)

prediction is the continuous line in Fig. 3.4. The cusp indicates the point where fracture commences. At shear rates beyond the cusp it is assumed that the fracture causes the external stress on the model to fall to zero and then a cycle of stress build-up recommences. The predicted stress is the average stress over one cycle. Most results of capillary testing do not conform to this prediction. However, the experimental points shown in Fig. 3.4 are an example of the behaviour typical of a number of compounds based on EPDM polymers which were previously considered to behave anomalously. The oscillation indicated in Fig. 3.4 is now seen to be due to the negative slope of the shear stress/shear rate curve combined with the ability of the extrusion rheometer to store energy in the rubber under compression in the barrel. A negative slope in the shear stress/shear strain characteristic would cause a dramatic increase in flow if the material were to be tested in a rheometer operating at a constant stress. This provides an explanation for the 'spurting' reported by Vinogradov *et al.* [8] in their work on highly homogeneous polymers.

The shear rate at which the cusp occurs is controlled by K and the fracture stress F. It would appear reasonable to expect that less homogeneous materials may be better represented by a range of

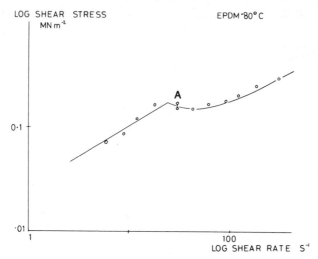

FIG. 3.4. ———, Typical prediction from the TMS model for shear stress versus shear rate during capillary flow; O, result from a capillary rheometer test on an EPDM formulation which conformed to this prediction; A, position and magnitude of shear stress oscillations noted during the test. (See Appendix 3.2.)

fracture stress values. Figure 3.5 shows a curve generated by including a range of five values for F. The use of a TMS model does, therefore, introduce concepts of fracture and homogeneity into the interpretation of the results from capillary rheometry.

There are, however, several additional factors which must be taken into account. The stresses in rubber in the capillary are greatest at the interface with the capillary wall and hence the nature of the fracture can be influenced by the surface of the wall. Wall slip, as reported by Worth and Parnaby [9], and fracture are probably related. Pressure will affect how much relaxation occurs after fracture and before the material coalesces again. In long capillaries the controlling layers experience large quantities of shear (several thousand units). As the properties of rubber change with shear, this leads to the material actually being tested being somewhat different to that tested in a short capillary. Temperature increases also need to be considered.

3.3.4   Die swell
One of the most important and elusive properties of polymers is extrusion length shrinkage or die swell, and an explanation for it has been the main objective for this work. Die swell depends on the time lapse after extrusion and so is difficult to measure experimentally [10]. Recovery over a period of time is more conveniently followed in a

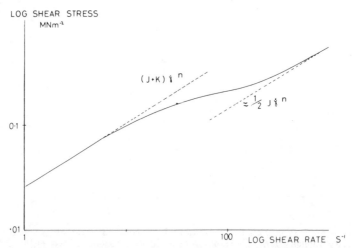

FIG. 3.5. Capillary flow shear stress versus shear rate prediction from the TMS model for a material having a range of fracture stresses. (See Appendix 3.2.)

shear recovery test. In this test a sample is placed in a cavity with a biconical rotor taken from a Monsanto rheometer. As shown in Fig. 3.6, after a sustained period of shear at a constant stress, the stress is removed and the progress of the recovery is followed. Figure 3.7 shows immediate recovery followed by time-dependent recovery.

Djiauw and Gent [2] examined the tensile recovery of unvulcanised rubber strips. In Fig. 3.8 their results have been re-plotted to be analogous to shear recovery results and the characteristic S-shaped curve was obtained. Gent noted that there was an immediate recovery followed by a time-dependent recovery.

The real value of the TMS model lies in its ability to predict immediate and time-dependent recoveries. It is difficult to visualise the effect of the equations and hence a qualitative view of the model is useful. The mechanism of recovery is easier to appreciate in a shear version of the model and its action is explained in Appendix 3.3.

Cotten [10] has shown the importance of providing a controlled environment for the extrudate so that time-dependent shrinkage can be followed. His original results were consistent with the later parts of S-shaped recovery curves, but at that time he was unable to make

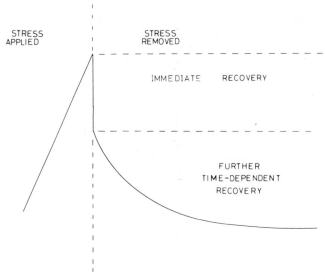

FIG. 3.6. Schematic representation of rotor movement with time during a typical shear recovery test.

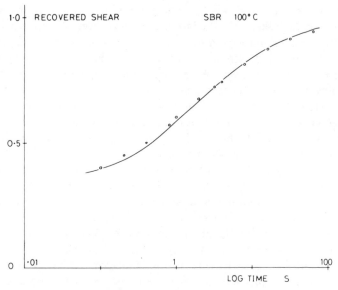

FIG. 3.7. ○, Result from a shear recovery test plotted against log(time); ———, a prediction of the TMS model with a suitable choice of parameters. (See Appendix 3.2.)

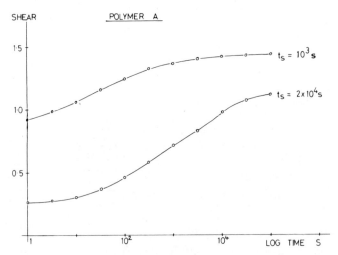

FIG. 3.8. Results from a tensile recovery test re-plotted to be comparable with shear recovery results.

measurements within one second of extrusion. It is an important test of the model to see if results at less than one second follow the first part of the S-shaped curve. Dr Cotten [11] carried out further experiments and his results are shown in Fig. 3.9. The experimental points at all times lie very well on a line predicted by the equations.

When extrusion shrinkage is plotted against extrusion pressure the shrinkage at low pressures increases with pressure, but eventually levels off. Figure 3.10 illustrates this. The interpretation provided by the model is that the elongational stress at entry is limited by fracture. Evidence of such fracture is shown in Fig. 3.11. The chamber of a Davenport rheometer was loaded with alternate discs of dark- and light-coloured rubber. At least 20 separate fragments of the dark-coloured rubber emerged, although only two layers reached the die entrance.

The full analysis found in the Appendices shows a very close relation between shear recovery and extrusion shrinkage after passage through a short die. However, the correlation is not complete as the elastic parameter E is involved in an additional way in extrusion shrinkage. Equations have also been developed to include a period of

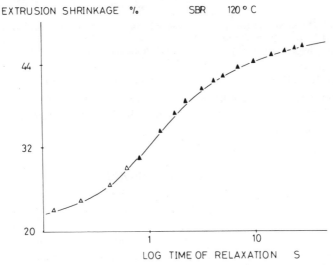

Fig. 3.9. Development of extrusion shrinkage plotted against the logarithm of the time after emergence from the die. ▲, △, Measurements made between 0·15 s and 30 s after extrusion; ———, a prediction of the TMS model with a suitable choice of parameters. (See Appendix 3.2.)

D. M. Turner, M. D. Moore and R. A. Smith

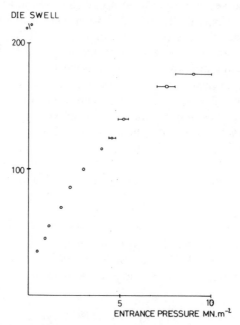

FIG. 3.10. Die swell plotted against entrance pressure from a very short die showing the limitation of die swell caused by high elongation rate at the die entry. The bars indicate the magnitude of pressure fluctuation during the test. (See Appendix 3.2.)

stress relaxation during passage through the die. The relationships between extrusion shrinkage and die length or speed of extrusion are very complex, which is why empirical methods have failed to reveal suitable equations.

3.4   CONCLUSIONS

The purpose this paper has been to show how a single model can explain the behaviour of unvulcanised rubber in a variety of situations, and we believe the results are sufficiently encouraging for it to be used as a foundation for many more research programmes. We believe that this concept will be most valuable in the following applications: (1) the choice of processing conditions for stock preparation for extruding and calendering; (2) the design of extrusion

FIG. 3.11. Illustration of fracture caused by elongational flow into the die. The upper section shows that only two of the dark rubber discs loaded into the chamber reached the die. The lower section shows consolidated extrudate where numerous fragments of dark rubber are visible.

dies and heads to be less sensitive to material variations; (3) the
development of quality controls which are most sensitive to the
parameters causing variations in production; (4) the development of
better specifications for the acceptance of raw polymers by rubber
processors; and (5) tests to investigate how different materials behave
in factory processes—aiding trouble shooting and optimisation of
process conditions.

REFERENCES

1. R. D. ANDREWS, A. V. TOBOLSKY and E. E. HANSON. *J. Appl. Phys.*, 1946, **17**, 352.
2. L. K. DJIAUW and A. N. GENT. *J. Poly. Sci.*, 1974, Symposia 48, 159.
3. P. G. NUTTING. *ASTM Proc.*, 1921, **21**, 1162.
4. G. R. COTTEN and B. B. BOONSTRA. *J. Appl. Poly. Sci.*, 1965, **9**, 3395.
5. R. A. DOVE, D. M. TURNER and T. MARTIN. (1977). In: *International Rubber Conference*, Vol. II, Brighton, UK, 16–20 May 1977.
6. J. FUNT. Imperial College, London, private discussion.
7. P. FREAKLEY and W. Y. WAN IDRIS. Loughborough University, private communication.
8. G. V. VINOGRADOV, N. I. INSAROVA, B. B. BOIKO and E. K. BORISENKOVA. *Poly. Eng. & Sci.*, 1972, **12**, 323.
9. R. A. WORTH, J. PARNABY and H. A. A. HELMY. *Poly. Eng. & Sci.*, 1977, **17**, 257.
10. G. R. COTTEN. (1977). In: *International Rubber Conference*, Vol. II, Brighton, UK, 16–20 May 1977.
11. G. R. COTTEN. Private communication.

Appendix 3.1

EQUATIONS DERIVED FROM THE TMS MODEL

NOTATION

E Modulus of elasticity for E/K network.
D Modulus of elasticity for D/J network.
K Viscous constant for E/K network.
J Viscous constant for D/J network.
n Power law index.

F Fracture stress.
S Extension.
$\dot{S}$ Extension rate.
Δt Time increment.
σ_E Stress on E/K network.
σ_D Stress on D/J network.
R Recovery.
R_0 Instantaneous recovery.
ΔR Recovery in time increment Δt.
L Extrusion shrinkage.
D_A Area die swell.
Y Yield following viscous flow.

NUMERICAL METHODS

Most of the equations described here cannot be solved analytically. Numerical methods were developed to allow the solutions to be computed by a desk-top programmable calculator.

The notation used in the following descriptions uses a dash (X') to represent the new value of a parameter X in a recurring loop.

(a) Elongation (see also section (h))

$$\sigma_E' = \sigma_E + E[\dot{S} - (1 + S)(\sigma_E/K)^{1/n}]\Delta t/(1 + S - \sigma_E/E)$$
$$\sigma_D' = \sigma_D + D[\dot{S} - (1 + S)\sigma_D/J)^{1/n}]\Delta t/(1 + S - \sigma_D/D)$$

where $S = Vt$ for constant extension rate tests, with V = pulling speed/sample length or $(1 + S) = \exp(\dot{e}t)$ for constant elongation rate tests with $\dot{e}$ = elongation rate.

In the interpretation of elongational test results, it helps a great deal to plot true stress against elongation ratio. This allows the elastic and viscous effects to be recognised at a glance.

(b) Creep

$\sigma = \sigma_0$ constant stress test or $\sigma = (1 + S)\sigma_0$ constant load test. Immediate extension

$$S_0 = \sigma/(E + D)$$

then

$$S'_K = S_K + \left(\frac{\sigma_E}{K}\right)^{1/n}(1 + S)\Delta t$$

$$S'_J = S_J + \left(\frac{\sigma_D}{J}\right)^{1/n}(1 + S)\Delta t$$

Effective spring stiffnesses also change:

$$E' = \frac{E}{(1 + S_K)}$$

$$D' = \frac{D}{(1 + S_J)}$$

Total extension is then $S = \sigma/(E' + D')$.

(c) Shear flow

There are two types of solution depending on whether or not the fracture stress on the E/K network is exceeded if the fracture stress is not exceeded, *i.e.*, $K\dot{\gamma}^n < F$

$$\sigma = (J + K)\dot{\gamma}^n$$

If the fracture stress is exceeded, *i.e.*, $K\dot{\gamma}^n > F$, after fracture the stresses are equal and opposite.

$$\sigma_{E0} = -\sigma_{D0} = (FD - EJ\dot{\gamma}^n)/(E + D)$$

These stresses increase with further shear until the fracture stress on the E/K network is reached.

$$\sigma'_E = \sigma_E + \left(\dot{\gamma} - \frac{\sigma_E}{K}\left|\frac{\sigma_E}{K}\right|^{(1-n)/n}\right)E\Delta t$$

$$\sigma'_D = \sigma_D + \left(\dot{\gamma} - \frac{\sigma_D}{J}\left|\frac{\sigma_D}{J}\right|^{(1-n)/n}\right)E\Delta t$$

The shear stress is then the time average of $(\sigma_E + \sigma_D)$ over the interval between fractures.

(d) Shear recovery

The previous analysis is used first to compute the initial stresses on the model. The immediate recovery is then given by

$$R_0 = (\sigma_E + \sigma_D)/(E + D)$$

This recovery changes the strains to

$$S_{E0} = \frac{\sigma_E}{E} - R_0$$

$$S_{D0} = \frac{\sigma_D}{D} - R_0$$

The time-dependent recovery is computed by cycling the following equations

$$\Delta R = \left[S_E - \left(\frac{ES_E}{K} \right)^{1/n} \Delta t \right] \left(\frac{E}{E + D} \right) + \left[S_D + \left(\frac{-DS_D}{J} \right)^{1/n} \Delta t \right] \left(\frac{D}{E + D} \right)$$

$$S'_E = S_E - \left(\frac{ES_E}{K} \right)^{1/n} \Delta t - \Delta R$$

$$S'_D = S_D + \left(\frac{-DS_0}{J} \right)^{1/n} \Delta t - \Delta R$$

where

$$R = R_0 + \Sigma \Delta R$$

(e) Stress relaxation in time t
$m = (n - 1)/n$ = relaxation constant

$$\sigma_E(t)^m = \sigma_E(0)^m - E^m/m \left(\frac{E}{K} \right)^{1/n} t$$

$$\sigma_D(t)^m = \sigma_D(0)^m - D^m/m \left(\frac{D}{J} \right)^{1/n} t$$

When computing residual stress resulting from the elongation rate at a die entrance, the values used for the parameters must be those for tension, not shear.

(f) Die swell and extrusion shrinkage
We can use the equations given in Section (d) for extensional recovery so long as the shrinkages R_0 and R are related to the extended length of the network in the die. We have used the value of $(1 + \sigma_E/E)$

as this length, valid for $D \gg E$, where σ_E is the stress on the E/F
network at the die exit.

Extrusion shrinkage is then

$$L = \frac{R}{(1 + \sigma_E/E)} \quad \text{for} \quad D \gg E$$

Area die swell D_A is related to extrusion shrinkage by

$$D_A = \frac{L}{1 - L}$$

These expressions can be readily evaluated on desk-top calculator
such as the HP 9815A.

(g) Effect of pressure on relaxation following model fracture

In the analysis given here it has been assumed that, following frac
ture, the external forces acting on the model are zero. Clearly, the
extent of this relaxation will depend on hydrostatic pressure—high
pressures will probably cause incomplete relaxation before the rubber
on either side of a fracture surface recombines.

(h) Effects of high extensions on flow formulae

The formula given in Section (a) is valid for one interpretation of the
model. A number of options are possible, which arise from changes in
the base dimensions of the sample as it becomes deformed. These
considerations are not too important when elongation ratios are less
than 2. Thus in the case of stress relaxation the simplest case was
used. Here as in applied stress situations the two networks act
independently so only a single network need be considered. The basic
case is:

$$\sigma = E(S - Y) = K(\Delta Y/\Delta t)^n \tag{1}$$

Y is the yield of the dashpot and removal of the external stress gives
$Y = S$, which constitutes the definition of Y when the original length
is 1.

The first refinement is that the extension rate applicable to the
generation of stress by viscous flow should be related to the in
stantaneous length of the sample.

$$\sigma = K[\Delta Y/(1 + S)\Delta t]^n \tag{2}$$

This expression was used to derive the equations in Section (a). An alternative which deserves some consideration is that

$$\sigma = K[\Delta Y/(1 + Y)\Delta t]^n \qquad (3)$$

Yielding of the sample creates extra length in the spring element, and so

$$\sigma = E(S - Y)/(1 + Y) \qquad (4)$$

Y relates to the unstrained length, but the quantity of flow may be concerned with the strained length, consequently

$$\Delta Y_1 = \Delta Y_2(1 + S - Y)/(1 + Y) \qquad (5)$$

ΔY_1 replaces ΔY in eqn (2); ΔY_2 gives $Y = \Sigma \Delta Y_2$.

As is pointed out in the paper, the model cannot be interpreted literally and it is difficult to decide whether eqns (3) and (5) should be used. Equation (5) should only be used in conjunction with eqns (2) and (4) but not (3).

Similar expressions apply to the D/J network.

Appendix 3.2

BASIC RUBBER FORMULATIONS, TEST CONDITIONS AND MODEL PARAMETERS RELEVANT TO EXPERIMENTS DESCRIBED IN THE TEXT

FIGURE 3.2

Rubber	100 phr	NR
	5 phr	Process oil
	100 phr	SRF Black
Test conditions	Instron tensile tester	
	Temperature: 80°C	
	Crosshead speed: 8·33 mm s^{-1}	
	Gauge length: 100 mm	
	Original cross-sectional area: 7·7 mm^2	

Model parameters $E = 78 \text{ kNm}^{-2}$ $K = 20 \text{ MNm}^{-2}\text{s}^n$
                       $D = 520 \text{ kNm}^{-2}$ $J = 545 \text{ kNm}^{-2}\text{s}^n$
                       $n = 0 \cdot 63$

FIGURE 3.3

Rubber   133 phr   OENR
                      50 phr   HAF Black
Test conditions   Constant stress creep apparatus
                      Room temperature
                      Applied stress = 50 kNm^{-2}
Model parameters

Energy input (MJm^{-3})	E (kNm^{-2})	K $(\text{kNm}^{-2}\,\text{s}^n)$	D (kNm^{-2})	J $(\text{kNm}^{-2}\,\text{s}^n)$	n
870	140	200	1000	60	0·39
1322	100	200	800	55	0·38
2365	90	200	600	50	0·37

FIGURE 3.4

Rubber   130 phr   OE EPDM
                      110 phr   FEF Black
                      50 phr   SRF Black
                      40 phr   Whiting
                      62 phr   Process oil
Test conditions   Davenport capillary rheometer
                      Temperature = 80°C
                      Die: 60-mm long; 2-mm diameter
Model parameters $E = 20 \text{ kNm}^{-2}$ $K = 6 \text{ kNm}^{-2}\,\text{s}^n$
                       $D = 500 \text{ kNm}^{-2}$ $J = 20 \text{ kNm}^{-2}\,\text{s}^n$
                       $n = 0 \cdot 5$ $F = 40 \text{ kNm}^{-2}$

FIGURE 3.5

The same model parameters as in Fig. 3.4 were used but with five values of fracture stress F.

FIGURE 3.7

Rubber	55 phr SBR 1712
	60 phr SBR 1500
	30 phr Oil
	77 phr N339 Black
Test conditions	TMS shear recovery tester
	Temperature = 100°C
	Applied shear stress = 136 kNm^{-2}
	Shear rate prior to recovery = 0·72 s^{-1}
Model parameters	$E = 65$ kNm^{-2} $K = 108$ kNm^{-2} s^n
	$D = 365$ kNm^{-2} $J = 80$ kNm^{-2} s^n
	$n = 0·3$ $F = 130$ kNm^{-2}

FIGURE 3.9

Rubber	Black-filled SBR
Test conditions	Monsanto capillary rheometer with laser die swell
	measuring apparatus
	Temperature = 120°C
Model parameters	$E = 58$ kNm^{-2} $K = 137$ kNm^{-2} s^n
	$D = 240$ kNm^{-2} $J = 72$ kNm^{-2} s^n
	$n = 0·46$ $F = 116$ kNm^{-2}
	Assumed shear rate = 10 s^{-1}

FIGURE 3.10

Rubber	100 phr BR
	56 phr SRF Black
	15 phr FEF Black
	27 phr Process oil
Test conditions	Davenport capillary rheometer
	Temperature = 100°C
	Die: 0·2-mm long; 2-mm diameter

Appendix 3.3

APPLICATION OF THE TMS MODEL
TO SHEAR RECOVERY

Figure 3.12 shows how a shear version of the TMS model produce
an instantaneous recovery. The diagram shows a platform with tw
sprung legs representing the D and E springs. On the bottom of eac
of these legs is a slider representing the viscous components J and K
The platform is pushed so that both the springs are deflected accord
ing to the stresses imposed by their respective viscous constants.
the stress is suddenly released, the platform will return to an equi

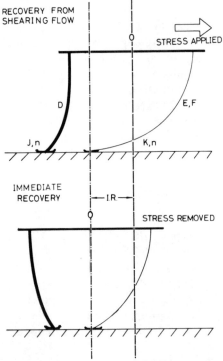

FIG. 3.12. Shear version of TMS model illustrating immediate recovery (I.R.).

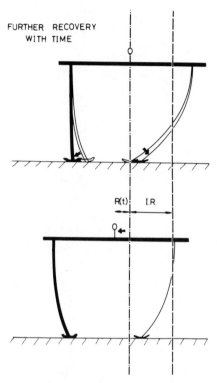

FIG. 3.13. Shear version of TMS model illustrating time-dependent recovery, R(t).
(I.R. = Immediate recovery.)

rium position where the forces on the two springs are equal and
opposed. This movement is the instantaneous recovery.

As seen in Fig. 3.13, the model is not yet at rest because of the
strains which still exist in the two springs. The two sliders will now
move according to their constants J and K. As spring D is so stiff, the
movement on D will cause a large reduction in the stress generated
by that spring so the platform will continue to move to the left to
re-establish the balance of forces. This movement will continue until
both springs are vertical. Appendix 3.1 contains equations describing
the action of the TMS model in shear recovery, based on this
mechanism.

Chapter 4

SOME ASPECTS OF THE TEAR STRENGTH
OF ELASTOMERS

A. N. GENT

4.1 INTRODUCTION

The strength of rubber is normally governed by its ability to dissipate the work expended in deformation. Striking evidence pointing to this conclusion was presented by Grosch, Harwood and Payne in 1966 [1]. They showed that the energy W_b expended per unit volume in stretching rubber to the breaking point was directly related to the energy W_d dissipated in irreversible processes when the material was deformed almost to break. Their results, shown in Fig. 4.1, follow the simple relation

$$W_b = 410 \ W_d^{2/3} \qquad (4.1)$$

where W_b and W_d are measured in J/m³. This relation was obeyed by a wide variety of rubber-like materials, ranging from the weakest to the toughest elastomers, *i.e.*, over the entire range of tensile strength. Thus, the exact mechanism of energy dissipation appears to be unimportant. For example, it may arise from strain-induced crystallisation, detachment from filler particles, fracture of weak bonds, or retarded segmental motion. Whatever the mechanism, the tensile strength of elastomers is governed by the extent to which they dissipate the energy of deformation.

The tear strength of elastomers also appears to be determined by dissipation processes. For example, the fracture energy T, defined as the work required to tear through unit area of the material, increases with increasing rate R of tearing and with decreasing temperature in the same way that any measure of viscous energy dissipation would

57

58 A. N. Gent

do, when a series of simple unfilled elastomers are compared [2]. Indeed, when the rates of tearing at different temperatures are converted into equivalent rates at the glass temperature, T_g, by means of the Williams, Landel and Ferry relationship [3] for the multiplying factor a_{T_g}, $\log a_{T_g} = -17 \cdot 5(T - T_g)/(52 + T - T_g)$, then the tear energies for different elastomers all superpose (Fig. 4.2) showing that the tear energy is governed by the rate of molecular sub-group motion for these simple visco-elastic solids [2].

What then is the tear strength of an elastomer under non-dissipative conditions, *i.e.*, at very low rates of tearing or at high temperatures or in the swollen state? This question is considered in the following section. Subsequently, some measurements of the strength of stretched elastomers are described.

4.2 THRESHOLD TEAR STRENGTH

A threshold tear strength for rubber was first pointed out by Lake and Lindley [4] in studies of crack growth under intermittent stress. If the

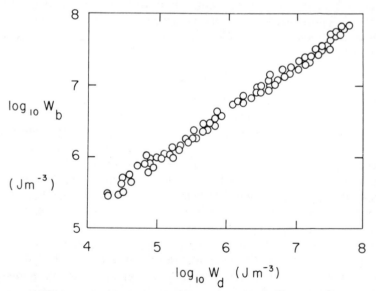

FIG. 4.1. Energy density W_b required for tensile rupture plotted against the energy density W_d dissipated in irreversible processes prior to rupture [1].

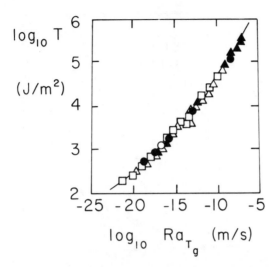

IG. 4.2. Tear energy T plotted against the rate R of tear propagation, adjusted to the lass temperature T_g by means of the WLF relationship. Results are shown for six SBR and NBR elastomers with T_g ranging from -80 to $-30°C$ [2].

nergy T applied for tearing did not exceed a critical value, about 0 J/m², then no crack growth occurred. This feature is shown in Fig. .3. Above the threshold value, denoted T_0, the amount of crack rowth Δc depended upon the energy T applied for tearing, and also pon the degree of energy dissipation of the rubber, being larger for nore elastic materials and smaller for more dissipative materials. Natural rubber, NR, is more dissipative than styrene–butadiene ubber (SBR) at high stresses because of strain-induced crystal-isation.) Thus, above T_0 the amount of tearing is in accord with the previously discussed dependence of strength upon dissipative pro-erties. However, T_0 itself is apparently independent of the dis-sipative properties.

Values of T_0 have since been determined directly, by tearing hot nd swollen elastomers at low rates [5, 6]. Under these conditions viscous effects are negligibly small and the tear strength approaches a ower limit which, after a correction is made in the case of swollen elastomers for the reduced number of molecules crossing the tear plane, is found to be between about 40 and 80 J/m². The exact value obtained for T_0 depends upon the degree of crosslinking of the elastomer, increasing somewhat as the molecular weight M_c between

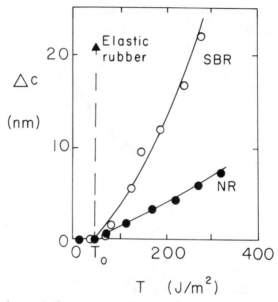

FIG. 4.3. Crack growth distance Δc per stress application plotted against the energy ?
available for fracture [7].

points of crosslinking is increased, *i.e.*, as the degree of crosslinking is
reduced, as shown in Fig. 4.4.

Lake and Thomas [7] have calculated the fracture energy T_0 of an
elastic network of hydrocarbon chain molecules from first principles,
obtaining the result

$$T_0 = Kn^{1/2} \qquad (4.2$$

where K is a constant involving the strength of a C–C bond, the
length and molecular weight of a chain sub-unit, Avogadro's Number
etc., and n is the number of chain sub-units between crosslinks. Thus
n is directly proportional to M_c. The full line in Fig. 4.4 is of the form
of eqn (4.2); it is seen to describe the experimental results reasonably
well at low values of M_c. Departures at large values of M_c are
undoubtedly due to chain-end effects when the value of M_c becomes
comparable to the original molecular weight of the elastomer before
crosslinking. Modification of the theory to take this effect into account has been carried out [7].

Thus, the theoretical treatment successfully accounts for the way in

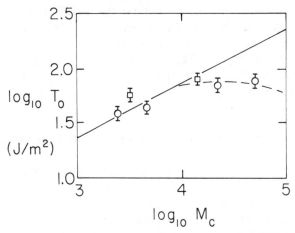

FIG. 4.4. Threshold tear strength T_0 plotted against the molecular weight M_c of network strands [6]. ———, Slope of 1/2 in accordance with eqn (4.2).

which the threshold tear strength depends upon the degree of cross-linking, over a wide range. Furthermore, the numerical values predicted are of the same order as those measured experimentally, being generally about 20 J/m² instead of about 50 J/m². This discrepancy is relatively small in view of many simplifying approximations made in the theoretical treatment and it must be concluded that the theory is basically correct.

Apart from the assumption that dissipation processes are absent, the theory also assumes that the tear is as sharp as possible for a highly-elastic solid, with a tip diameter in the unstrained state of the order of the end-to-end length of a network strand of molecular weight M_c, about 100–200 Å. In contrast, under normal tearing conditions, the tear tip is found to be quite blunt, having an effective diameter of 0·1 mm or more [8]. When reinforcing fillers are present a still greater tear bluntness is encountered because the tear deviates from a straight path and turns into a direction parallel to the principal tensile stress [9]. Tearing is arrested by this lateral deviation until a new tear forms and runs ahead for some distance before it also undergoes a similar deviation. This characteristic intermittent or 'knotty' tearing on a scale of some millimetres is associated with a particularly high resistance to tearing. Thus, a general correlation is found between the natural bluntness of the tear tip and tear strength, in addition to the previously-discussed correlation with energy-dissipation processes. Indeed, the blunter the tear tip the greater is the volume of material that undergoes large deformations and hence the greater the expected amount of energy lost owing to mechanical hysteresis during tearing. The two contributions to tear strength are thus intimately related in some instances, although they are in principle independent.

4.3   TEAR STRENGTH OF STRETCHED RUBBER

We now turn to some recent measurements of the effect of a large tensile strain upon the tear strength of reinforced elastomers. Two quite different experiments have been carried out—one in which the sample is stretched and released before tearing, and one in which the stretched sample is torn in such a way that the tear runs in the direction of extension [10]. The first experiments revealed that in most cases using filled and unfilled SBR materials, unfilled NR, and a thermoplastic elastomer Kraton 1101 (Shell Chemical Co.), large extensions had no significant effect on the tear strength after release. Results for a carbon-black-filled SBR material are plotted in Fig. 4.5 against the amount of pre-stretch for a tear running either parallel or perpendicular to the direction of pre-stretch. There is little difference between them and the original value.

For a carbon-black-filled NR material, however, the effects of prior stretching were dramatic (Fig. 4.6). In this case the tear strength for a tear running in the direction of pre-stretch fell to about one-sixth of

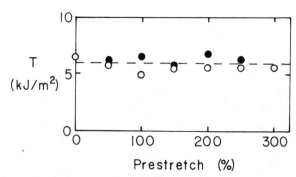

FIG. 4.5. Tear strength of carbon-black-reinforced SBR plotted against the amount of prior extension [10]. ○, Tearing parallel to the pre-stretch direction; ●, tearing perpendicular to the pre-stretch direction.

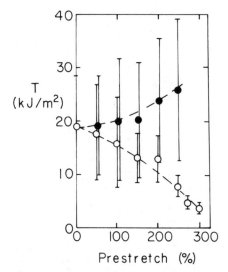

FIG. 4.6. Tear strength of carbon-black-reinforced NR plotted against the amount of prior extension [10]. ○, Tearing parallel to the pre-stretch direction; ●, tearing perpendicular to the pre-stretch direction.

the original value and the torn surface became relatively smooth, whereas the tear strength for a tear running perpendicular to the pre-stretch direction rose somewhat and retained the characteristic large fluctuations associated with intermittent, knotty, tearing. A memory of previous stretching is therefore strongly evident in this material in contrast to all the others examined. It must be attributed to the combination of a strain-crystallising elastomer with a reinforcing filler, and suggests that crystallisation brings about a segregation of filler particles so that weak paths are left in the elastomer subsequently, when the strain is released and the crystallites melt. Thus, in this case the reduced tear strength appears to be a consequence of a sharper tear tip, rather than any substantial change in dissipative properties. A similar conclusion has been reached previously [11].

When the testpiece was held stretched during tearing, the tear strength fell sharply for all the materials examined. The results for carbon-black-filled SBR and NR materials and for Kraton 1101 are plotted in Fig. 4.7 against the imposed extension e_1 for a tear running in the direction of extension. (No measurements were made for a tear running perpendicular to the extension direction.) The effect is seen to be large in all cases, although far from equal. For the filled SBR

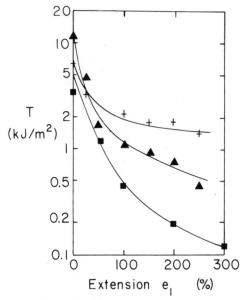

F IG. 4.7. Tear strength of stretched elastomers plotted against the imposed extension, e_1 [10]. +, Carbon-black-reinforced SBR; ▲, carbon-black-reinforced NR; ■, Kraton 1101 (Shell Chemical Co.).

elastomer the effect is smallest, the tear strength in the highly-stretched state being about 1500 J/m^2, *i.e.*, about 25 per cent of the tear strength in the unstretched state. For the filled NR material, the tear strength fell to about 500 J/m^2, only about 5 per cent of the original value, and for Kraton 1101 it fell to about 120 J/m^2, close to the threshold strength of elastomeric networks and far below the value in the unstretched state.

These striking changes in tear strength also appear to reflect easy fracture paths parallel to the extension direction and a consequent increase in sharpness of the tear tip rather than major changes in dissipative properties [11]. Indeed, other work has shown that the dissipation of energy for perpendicular deformations, occasioned by tearing stresses in the present instance, is not much affected by a maintained extension [12].

Whatever the cause, large decreases in tear strength in the direction of an imposed extension appear to be a general feature of elastomers although the effect is not equal for different elastomers. This observation indicates that a (variable) degree of strength anisotropy exists

highly-stretched rubber. Thus, if a relatively small tensile stress is
set up at right angles to the principal tension, the rubber will split
apart readily. This may be the origin of two characteristic features of
strong elastomeric systems: knotty tearing, when the tear deviates
into the direction of the principal stress, *i.e.*, splitting occurs in this
direction; and microscopic splitting on the scale of a few micrometres
observed at the tear tip, responsible for a characteristic roughness of
the torn surface. In both cases, the easy splitting of highly-stretched
rubber may be a direct source of strength by causing the tear tip to
become blunter.

ACKNOWLEDGEMENTS

This review was prepared in the course of a program of research on
fracture supported by a Research Grant from the Engineering
Division of the National Science Foundation. It is based on recent
research carried out in this program [6, 10–12].

REFERENCES

1. K. A. GROSCH, J. A. C. HARWOOD and A. R. PAYNE. *Nature*, 1966, **212**, 497.
2. L. MULLINS. *Trans. Instn Rubb. Ind.*, 1959, **35**, 213.
3. J. D. FERRY. (1970). *Viscoelastic Properties of Polymers*, 2nd edn, New York: John Wiley & Sons.
4. G. J. LAKE and P. B. LINDLEY. *J. Appl. Poly. Sci.*, 1965, **9**, 1233.
5. H. K. MUELLER and W. G. KNAUSS. *Trans. Soc. Rheol.*, 1971, **15**, 217.
6. A. AHAGON and A. N. GENT. *J. Poly. Sci. Poly. (Phys. Edn)*, 1975, **13**, 1903.
7. G. J. LAKE and A. G. THOMAS. *Proc. Roy. Soc. (London)*, 1967, **A300**, 108.
8. H. W. GREENSMITH, L. MULLINS and A. G. THOMAS. (1963). In: *The Chemistry and Physics of Rubberlike Substances* (Ed. L. BATEMAN), Ch. 10. New York: John Wiley & Sons.
9. H. W. GREENSMITH. *J. Poly. Sci.*, 1956, **21**, 175.
10. A. N. GENT and H. J. KIM. *Rubb. Chem. Tech.*, 1978, in press.
11. A. N. GENT and A. W. HENRY. (1967). In: *Proceedings of the International Rubber Conference, 1967*, pp. 193–204. London: Maclaren.
12. J. M. CHARRIER and A. N. GENT. (1975). *Les Interactions entre les Elastomeres et les Surfaces Solides ayant une Action Renforcante*, Colloques Internationaux No. 231, pp. 205–211. Paris, France: CNRS.

Chapter 5

TEAR BEHAVIOUR OF RUBBERS OVER A WIDE RANGE OF RATES

A. KADIR and A. G. THOMAS

5.1 INTRODUCTION

The tear behaviour of rubbers has been shown [1–4] to be conveniently described in terms of the tearing energy (T) versus rate of tear (r) relation. The quantity T is defined by

$$T = \frac{-1}{h}\left(\frac{\partial U}{\partial c}\right)_l$$

where U is the elastically stored energy in the testpiece of thickness h, c the length of the crack and l is the overall length so that the partial derivative indicates that the external forces do not move. When expressed in this way, the tear behaviour is found generally to be independent of the testpiece shape and T is thus a fundamental strength characteristic of the material. An apparent anomaly was noted however [4] when comparisons were made between results from a pure shear test piece (Fig. 5.1) and those from a 'trousers' test piece in the region where the tear behaviour was irregular and showed a 'stick-slip' character. This was ascribed to the fact that with the trouser testpiece the rate fluctuated substantially as did also the tearing force, and average values could therefore be misleading. The pure shear testpiece avoids certain of these complications as the T value is held essentially constant, independent of r, and the latter may be measured and any fluctuations noted. This constancy of T will hold as long as inertial effects can be neglected, which means that r must not approach the velocity of elastic waves in the rubbers, which is typically about 50 m s^{-1}. Some earlier work [5] indicated that at high

67

T values a significant fraction, about 0·3, of this velocity could be attained.

The present work extends the range of crack propagation rates to substantially higher values than have been systematically studied in the past. For the higher rates a high speed camera was used, with framing rates up to a few thousand per second.

5.2 EXPERIMENTAL

The materials used were unfilled compounds of natural rubber (NR), butadiene acrylonitrile rubber (NBR), styrene butadiene (SBR) and a polybutadiene (BR) with the formulations given in the appendix. The stress–strain behaviour in pure shear was first determined for each material. A sheet of the material to be studied, about 1 mm thick, was marked with a suitable grid of ink lines and then clamped in grips about 25 cm long, their initial separation being variable and usually in the range of 1 to 4 cm. For this pure shear testpiece (Fig. 5.1) the tearing energy is given by

$$T = Wl_0 \qquad (5.1)$$

where W is the stored elastic energy per unit volume in the pure shear region and l_0 the initial length of the testpiece. The testpiece was then deformed and the strain in the pure shear region measured, from which W was obtained. A cut was then inserted in one edge and the movement of the crack tip recorded, if necessary by using the high speed camera.

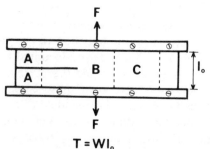

FIG. 5.1. Pure shear testpiece. A, relaxed region; B, region of complex strain; C, pure shear region. l_0, unstrained length of testpiece.

5.3   RESULTS AND DISCUSSION

The relation between T and the crack propagation rate r is indicated qualitatively in Fig. 5.2 for non-crystallising rubbers. The behaviour of such materials falls into three regions, indicated in the figure. At low rates, less than about 10^{-2} cm s^{-1} for SBR, the tearing is fairly steady and a typical crack length against time plot is shown in Fig. 5.3(a). There are significant changes in instantaneous rate but the average rate can nevertheless be estimated with fair confidence. The appearance of the torn surface is rough, and is shown in Fig. 5.4(a). In the region of stick-slip tearing, Fig. 5.3(b) indicates very large changes in instantaneous rate. Here the average value has little significance as it will depend strongly on how much time is spent in a stick or slip mode and this appears to be governed largely by chance. In contrast, the high speed (>20 m s^{-1}) smooth region shows a remarkably constant rate (Fig. 5.3(c)). The appearance of the surfaces in these regions is illustrated in Figs. 5.4(a)(b) and (c). The high speed, constant rate, region shows a very smooth surface which is reminiscent of a glassy fracture, although examination does show certain features such as marks near the centre of the torn surface which have an approximately paraboloidal shape. Although these are reminiscent of marks which have been observed many times on the fracture surfaces of glassy polymers [6], unlike these they point in the *same* direction as the propagating crack. The usual explanation of the paraboloidal

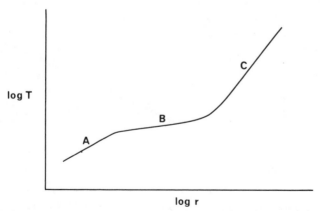

Fig. 5.2. Schematic T against rate diagram. A, rough tearing; B, stick-slip tearing; C, smooth tearing.

70   A. Kadir and A. G. Thomas

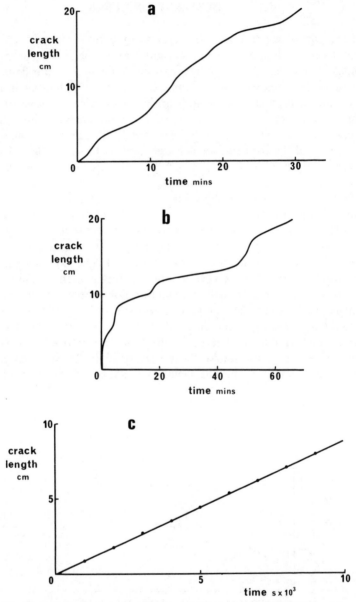

FIG. 5.3. (a) crack length c against time t plot for tearing of SBR in the rough region indicated in Fig. 5.2 ($T = 2 \cdot 2\,\text{kNm}^{-1}$); (b) c against t for stick-slip region for NBR ($T = 4 \cdot 0\,\text{kNm}^{-1}$); (c) c against t for smooth region for NR ($T = 14 \cdot 0\,\text{kNm}^{-1}$).

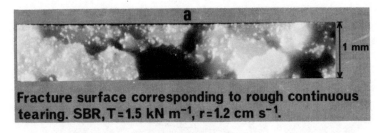

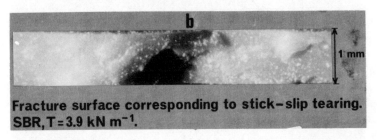

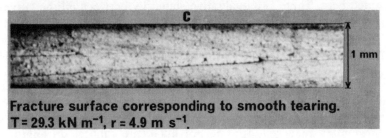

Fig. 5.4. (a) surface in rough tear region; (b) surface in stick-slip region; (c) surface in smooth tear region. The testpieces are about 1 mm thick.

marks on fracture surfaces is that they are produced by the inter-action of the main crack front with subsidiary fracture fronts initiated from flaws just ahead of the crack. To explain the present observations along these lines it is necessary to assume that the subsidiary fractures are initiated ahead of the main front but propagate more slowly and so are overtaken by this front.

Figure 5.5 shows the tearing energy against rate relation for the four materials studied. Each material was studied using various lengths l_0 of the testpiece, and within the reproducibility of the measurements these different lengths give equivalent results. This confirms the validity of the energetic (fracture mechanics) approach

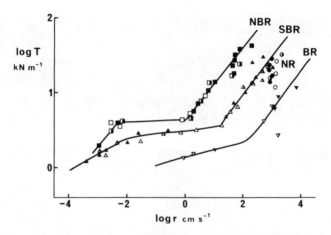

FIG. 5.5. Tearing energy against rate relations for NBR, □, ◧, ■; SBR, △, ▲, ▲; NR, ○, ◑, ●; BR, ▽, ▼, ▼. The l_0 values are 2·0, 3·0, 4·0 cm, respectively.

adopted here. There is a qualitative similarity between all the curves. The high speed regions show a correlation with the glass transition temperatures of the rubbers as has been found before for relatively low speed tests [7]. The rates attained do not in general approach that of the velocity of shear waves in rubber, about $50 \, \text{m s}^{-1}$, and this shows that even in this region where the fracture surfaces are smooth and 'glassy' in appearance the intrinsic strength of the material, and thus its visco-elastic behaviour, still controls the tearing rate. The highest rates recorded, for the polybutadiene rubber, do approach the wave velocity, but even here there is no indication from the data that this limit is having a substantial influence on the results.

At T values below about $1 \, \text{kNm}^{-1}$ the natural rubber compound does not tear continuously. This is consistent with previous work [2] and indicates the presence of strain induced crystallisation at the tip of the crack. At higher T values the natural rubber tears at a rate appropriate to its glass transition temperature, that is, intermediate between SBR and BR. Thus at the high rates of loading produced at the tip by these rates of propagation, about $10 \, \text{m s}^{-1}$, the rubber cannot apparently crystallise.

The influence of visco-elasticity on strength has been demonstrated by showing that the equivalence between changes in rate and temperature of test follows the well-known Williams–Landel–Ferry (WLF) relation [7, 8]. The results shown in Fig. 5.5 were obtained at

20°C, but the rubbers had widely different glass transition temperature (T_g) values (BR, −112°C; NR, −73°C; SBR, −63°C; NBR, −31°C). An attempt was made therefore to see if the WLF relation would produce a single master curve of the T against rate relation for the various rubbers in the smooth tearing region.

The usual form of WLF relation is

$$\log a_T = -\frac{8 \cdot 86 \, (T - T_s)}{101 \cdot 6 + T - T_s} \tag{5.2}$$

where a_T is the factor by which the rate must be multiplied to give a master curve, and T_s is a reference temperature characteristic of each rubber. It has been found that T_s can usually be taken as $T_g + 50°C$. In the present case however it was found possible to obtain a master curve only by taking T_s as $T_g + 20°C$ (Fig. 5.6).

This somewhat unexpected finding was checked by further experiments on the NBR rubber at 70°C, so that the effect of changing the test temperature rather than the glass transition temperature could be examined. The results, transformed in the same way as before, are shown in Fig. 5.7. It was again found necessary to set $T_s = T_g + 20°C$ for the transformation to be effected. This is supporting evidence for the correctness of this identification of T_s.

This conclusion is consistent with a depression in the glass transition temperature of the material ahead of the tip by about 30°C. It is interesting to compare this suggestion with a theory proposed by Gent [9] that crazing in glassy plastics may be due to the hydrostatic com-

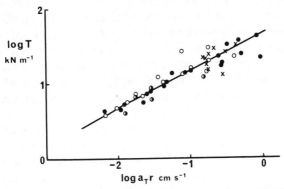

FIG. 5.6. Composite curve for smooth tearing, ● NBR, ○ SBR, × NR, ◑ BR. The shift factor a_T is given in the text.

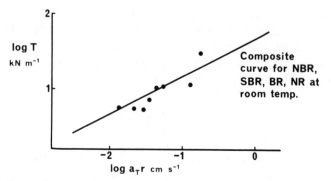

FIG. 5.7. Composite curve for smooth tearing with results for NBR at 70°C. The full curve is the same as that of Fig. 5.6.

ponent of the tensile stress field around a flaw depressing the glass transition enough to cause devitrification of the polymer. A similar mechanism may well be applicable here as tensile stresses around the tip will be high (perhaps of the order required to break chemical bonds) and their hydrostatic component would be expected to produce a significant depression in T_g.

The abrupt break in the $T - r$ relation as the T value is reduced correlates with the development of a rough tip. The importance of the detailed structure of the tip of the growing tear has been emphasised before [2, 3]. During measurements under repeated stressing of gum natural rubber compounds the crack growth rate has been observed to decrease by about a factor of ten as the tip grew from an initially smooth, razor cut, tip to a rough tip [10]. In this case it was suggested that the rough tip could be considered to consist of a number of sharp tips adjacent to each other so that the tearing energy was effectively shared between them. The number of such tips was estimated to be about 3, this being consistent with both the reduction in crack growth rate and the increase in the catastrophic tear strength when tearing from a rough compared with a razor cut tip. It is tempting to think that a similar effect may be operative in the present measurements. Some further information is given by a closer analysis of the crack length against time relations of the sort shown in Fig. 5.3(a–c). A histogram can be constructed showing the relative frequency distribution of instantaneous tearing rates, obtained by taking the instantaneous slopes of the crack length against time curves. In the high speed (smooth) region there is little variation in the rate, but in the stick-slip and rough regions histograms such as those shown in Figs. 5.8(a) and (b) are obtained. These show that the variations about the average rates are not randomly distributed but tend to cluster around distinct values, which may be quite widely separated. If the model of a rough tip consisting of several elementary sharp tips is accepted, those results can be understood in terms of abrupt random changes in the number of these tips as the tear proceeds.

The influence of the roughness of the tip can be illustrated by the effect of testpiece thickness on the tear behaviour. The roughness developed on the surface is of the order of a few tenths of a millimetre in scale. If the testpiece thickness is also of this order, then it might be expected that the scale of the roughness, and hence the tear behaviour, would be influenced by the thickness. Results for SBR for two thicknesses are shown in Fig. 5.9. In the smooth tear region no effect of thickness is found, but a significant effect is seen in the rough and stick-slip regions. This is consistent with expectations as

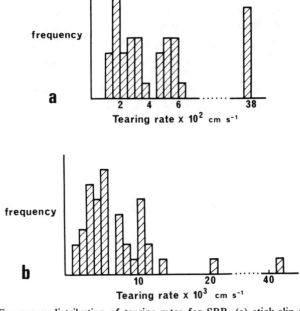

FIG. 5.8. Frequency distribution of tearing rates for SBR, (a) stick-slip region, $T = 2\cdot9\,\text{kNm}^{-1}$; (b) rough region, $T = 2\cdot2\,\text{kNm}^{-1}$.

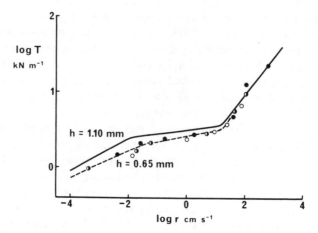

FIG. 5.9. Tearing energy against rate of tear for SBR showing the effects of thickness h. Testpieces of thickness 0·65 mm with various l_0 values were used: ○, 2·0 cm; ◑, 3·0 cm; ●, 4·0 cm. The full curve is that from Fig. 5.5 with $h = 1·1$ mm.

the smooth tear region presumably produces as sharp a tip as possible, but the development of the full potential roughness is inhibited by the thickness limitation in the rough and stick-slip tear regions. Figure 5.10 illustrates the effect in these latter regions more fully. The thickness has most influence when it is a little less than 1 mm, consistent with the observed magnitude of the roughness.

The problem still remains of why there is the abrupt change in tear characteristics and surface appearance. A possible explanation can be put forward along the following lines. It is known that under hydro-

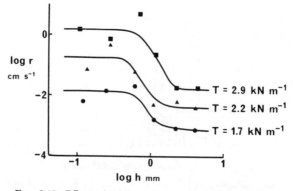

FIG. 5.10. Effect of thickness on tear behaviour of SBR.

tatic tension rubber can cavitate [11], and that the stress necessary is
f the order of the Young's modulus of the material. Just ahead of the
rack, high tensile stresses must be developed locally to rupture the
rimary chemical bonds. These stresses will be orders of magnitude
reater than that required to produce tensile failure in conventional
ensile tests because of the influence of flaws in the latter case. A
recise estimate of such a theoretical strength is difficult, but con-
ideration of bond strengths in a similar manner to that described by
ake and Thomas [12] suggests that stresses referred to the un-
leformed cross-section of the order of $10^9 \, \mathrm{Nm^{-2}}$ or more might be
eached. The hydrostatic component of such a stress would be
idequate to produce cavitation in a rubber of conventional modulus
$10^6 \, \mathrm{Nm^{-2}}$), and it is suggested that this may be the cause of the
oughness developed around the tip. However, when the crack is
propagating rapidly the rate of extension ahead of the tip can be
extremely high. For a tip whose diameter is governed by the chain
end-to-end distance ($\sim 10^{-6} \, \mathrm{cm}$) and for a crack propagation rate of
20 cm s^{-1} (the transitional rate for SBR) the rate of extension of the
rubber will be of the order of $10^8 \, \mathrm{s^{-1}}$. This will be rapid enough to
increase the modulus very substantially, and as the glassy modulus of
polymers is about $3 \times 10^9 \, \mathrm{Nm^{-2}}$ it is suggested that rupture of the
chains can now occur before cavitation takes place. This mechanism
is based on the visco-elastic increase in modulus with rate of extension and thus the correlation of the transition tearing rate with T_g
shown in Fig. 5.5, is to be expected. Figure 5.5 also indicates that the
transition tearing energy is not very sensitive to the nature of the
rubber. This again might be expected on the theory proposed as the
intrinsic strength of the materials, being essentially governed by the
carbon–carbon bonds, will not vary much and the transition points
will thus be at equivalent visco-elastic conditions for the rubber at the
tip and producing therefore similar strengths. The apparent reduction
in T_g, deduced from the rate–temperature transformation and discussed
earlier, is suggestive of a substantial hydrostatic tensile stress near the
tip, which is consistent with the proposed mechanism.

REFERENCES

1. R. S. RIVLIN and A. G. THOMAS. *J. Polym. Sci.*, 1953, **10**, 291.
2. H. W. GREENSMITH and A. G. THOMAS. *J. Polym. Sci.*, 1955, **18**, 177.

Chapter 6

MECHANICAL BEHAVIOUR OF ELASTOMERS IN FOOTWEAR

R. E. WHITTAKER

6.1 INTRODUCTION

On joining SATRA in 1968, Bob Payne was anxious to apply some of the published theories [1–4] of rubber reinforcement to typical materials being used in the footwear industry. In the late 1960s and early 1970s the use of cellular polyurethanes either as upper or soling materials had begun to develop and, together with the author, studies were undertaken using the earlier work on vulcanised rubbers in order to explain the high strength of these polymers.

Dr Payne's interest in the engineering of rubbers was also used as part of the basis for SATRA's shoe engineering work which has developed during the last few years, the concept being to consider a finished shoe as a structure which needs to be properly engineered. By measuring the strains and stresses to which the shoe is subjected in wear, specifications in terms of physical properties can be developed for the component parts, particularly on fashion shoes. The combination of the understanding of the physical properties of cellular polyurethane elastomers together with measurements of deformation on shoes in wear from SATRA's shoe engineering work, has enabled a full understanding to be obtained of recent failures in wear of cellular polyurethane soles and upper materials.

6.2 CELLULAR POLYURETHANES

Cellular polyurethanes have been used in the footwear industry in recent years [5–8] in closed-cell form as soling materials and as

79

open-cell foams in upper materials. The majority of polyurethanes ar
of a linear type and are based on a polyester polyurethane. Polyether
based compounds can be used but they do not have the high strength
and good flexing performance obtained with the polyester type
Generally, 4,4'-diphenylmethane diisocyanate (MDI) is used in pref
erence to other isocyanates such as naphthalene diisocyanate (NDI
or toluylene diisocyanate (TDI) as it is reasonably priced and ha
medium reactivity. It is also available in the pure state and therefor
allows products of high consistency and good stability to b
manufactured. These materials generally have high strength, abrasion
and cut growth resistance even in cellular form and a tensile curve fo
a polyurethane foam material is shown in Fig. 6.1. The tensile stres
for the foam is based on the cross-sectional area of the rubber
including the holes. The tensile stress–strain curve for the solic
polyurethane which was obtained by dissolving the foam in a suitable
solvent and recasting the sheet is also shown in Fig. 6.1 as well as the
stress–strain curve for a typical natural rubber vulcanisate from
earlier investigations [2, 3].

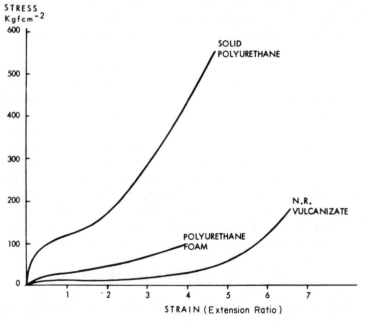

FIG. 6.1. Typical stress–strain curves for solid and foam polyurethane and solid NF
vulcanisate.

It would appear that the initial modulus of the polyurethane foam is igher although the actual tensile strength is lower than a solid NR ulcanisate. The modulus of the solid polyurethane is extremely high hen compared with the corresponding foam and its tensile strength considerably in excess of that found in the natural rubber vulanisate.

Some years ago, Gent and Thomas [9] derived a theory for xpressing the mechanical properties of a latex foam rubber in terms f the corresponding solid material. The model they used consisted of cubical array of struts. The author [10] has previously shown that is model can be applied successfully to the mechanical properties of icrocellular polyurethane foams.

The high strength of the polyurethane foam is therefore due to the igh strength of the solid polyurethane material, the ratio between the echanical properties of the foam and solid materials being the same s between the solid and foam vulcanised rubbers of the same ensity.

.2.1   Failure properties

revious studies [2–4] have shown that a useful measure of the trength of a polymer is the toughness or energy input to break as it ombines both the contributions due to stress and strain at break. The ariation of energy input to break with temperature for both the foam nd solid polyurethanes at a strain rate of 2·2 per min is shown in Fig. 2. The energy input to break values for both the foam and solid olyurethanes remain fairly high and parallel up to approximately 60°C when the failure values drop quite markedly. In order to idicate the high strength and temperature stability of poromeric olyurethanes, results for normal styrene–butadiene rubber (SBR) ith 0 and 30 phr HAF carbon black from earlier investigations [2, 4] re also shown for comparison in Fig. 6.2. It is clearly seen that even e polyurethane foam has higher strength properties over the majrity of the temperature range considered than has the solid reinorced rubber vulcanisate.

.2.2   Physical structure of polyurethane elastomers

order to explain some of the anomalous features of polyurethanes ompared to vulcanised rubbers, it is necessary to consider the hysical structure of these materials. A number of investigators 1–15] in recent years have shown that polyurethane elastomers

R. E. Whittaker

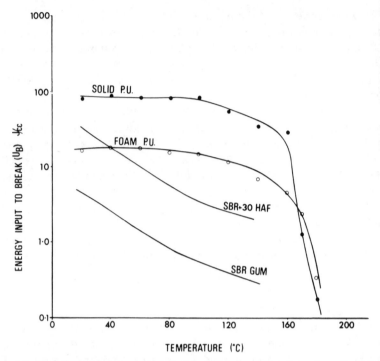

FIG. 6.2. Variation of input energy to break with temperature for polyurethane foam and solid materials compared with filled and unfilled SBR vulcanisates.

consist of alternating hard and soft segments as shown in Fig. 6. The soft segments are formed from the linear polyether or polyester chain segments which are about 100–200 Å long and at service temperatures are sufficiently high above their glass transition temperature, or in the case of crystallisable soft segments above the melting temperature, to give the material an extensibility of sever hundred per cent. The hard segments are approximately 25 Å length and originate from a diisocyanate and a chain extender crosslinking agent (e.g. diol or diamine) and hence contain urethane urea groups. These hard segments are at service temperatures below their second-order glass transition temperature and cause physic crosslinking by hydrogen bonds and other intermolecular forces between the segments. This prevents the material from flowing so th elasticity is maintained.

From X-ray and differential thermal analysis (DTA) measurement

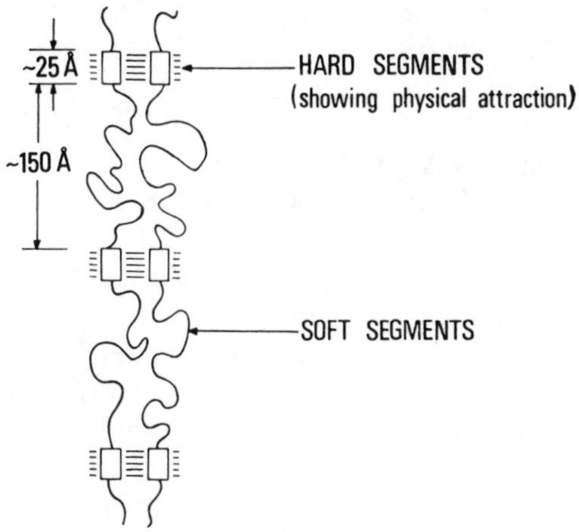

FIG. 6.3. Diagrammatic representation of typical structure of a polyurethane elastomer.

16–20] it has been shown that two major glass transition temperatures occur in polyurethane elastomers. The first occurs around
-20°C and is due to the onset of rotation in the flexible polyether or
polyester chain, whereas the second transition at about 160°C is due
to the dissociation of the inter-urethane hydrogen bonding.

The most interesting feature of the physical structure of polyurethane elastomers is their segmented structure. The hard urethane
segments presumably act as filler particles within the polyether or
polyester rubber matrix. It is well known that the introduction of a
filler such as carbon black into an amorphous vulcanised rubber
increases its hardness, strength, hysteresis and abrasion resistance.
The large increase in these properties in the case of polyurethane
must be due, however, to the very small size of the hard urethane
segment of the chain. The normal carbon blacks used in rubbers (e.g.
SRF, ISAF, HAF etc.) are about 300 Å in diameter whereas in the
case of polyurethane the hard segments are a factor of ten smaller
than this. It is well known that the properties of filled vulcanised
rubbers improve as the particle size of the filler is reduced as there is
more particle surface area to which the rubber chains can adhere. In
the case of polyurethanes the very small size of the filler particle
causes the very high strength and abrasion resistance.

The other interesting feature of polyurethane elastomers is that the
filler particles (the hard segments) form a proper molecular arrange-

ment. The filler particles are therefore well dispersed in the polyeste
or polyether rubber matrix, a secondary factor which contributes t(
their good mechanical properties.

6.2.3 Cut growth resistance

One of the most important properties for polyurethanes used i
footwear is that they have a high resistance to cut growth. In order t(
determine the cut growth properties of microcellular polyurethanes
some measurements were undertaken on the foam polyurethane
used in these materials. The measurements were based on the tearin
energy theory which was developed for vulcanised rubbers so that
direct comparison with the cut growth properties of rubber could b(
obtained.

The concept of 'tearing energy' for polymeric materials was ori
ginally developed by Rivlin and Thomas [21] to describe the tea
behaviour of vulcanised rubbers and is an extension of the classica
theory for the strength properties of glass which was developed b
Griffith [22] in 1920. The parameter, tearing energy, serves as
measure of the consequences of the high stress concentrations whic
surround the tip of a flaw in a highly strained piece of rubber [23–25]
It is defined for a strained testpiece containing a crack as

$$T = \partial U / \partial A)_e \tag{6.1}$$

where U is the total elastically stored energy in a testpiece and A
the area of the two sides of the cut surface. The derivative must b(
taken under conditions that the applied forces do not move and henc(
do no work. The suffix e denotes that the differentiation is carried ou
at constant deformation. It thus represents the rate of release of strai
energy as the crack propagates and can therefore be considered as th
energy available to drive the crack through the material. It has bee
found that if tear or crack-growth measurements are expressed i
terms of T, the results obtained from testpieces of different shape
can be correlated [23–26].

For a testpiece in the form of a strip with a small cut of length c i
one edge, deformed in simple extension as in the measurement
on polyurethanes, the tearing energy [26] is given by

$$T = 2KWc \tag{6.2}$$

where W is the strain energy density in the bulk of the testpiece (i.(
away from the cut) and K is a slowly varying function of strain whic

as been determined empirically for strains up to 200 per cent by ireensmith [26].

It has been found for vulcanised rubbers [23–25] that a minimum alue of tearing energy (T_0) exists below which there is no mechanical ut growth and hence this defines a fatigue limit for repeated stressing •elow which the life can be indefinite in the absence of chemical ffects.

The cut growth experiments on polyurethanes were performed on ensile strips of approximate dimensions 15×25 cm and about 2 mm hick. A cut about 0·5 mm long was made in the centre of one edge of he sample by use of a razor blade and the testpiece was then lamped into position on a repeated extension machine and extended o a suitable strain and cycled.

The rate of cut growth (dc/dn) was determined from the difference n cut length divided by the number of cycles between the two eadings. This rate was then referred to the tearing energy calculated rom the average of the two cut lengths and the corresponding $2KW$ ·alue at the maximum initial strain of the cycle. The test was stopped vhen the cut reached about 20 per cent of the testpiece width as the heory is unapplicable above this cut width. It was possible, however, o cover a decade of tearing energy values with one testpiece. A lifferent range of T was covered by cycling another sample to a lifferent maximum strain, hence changing $2KW$.

The variation of the rate of cut growth (dc/dn) with tearing energy 'or the microporous polyurethane foam is shown in Fig. 6.4. The owest recorded value of tearing energy at which some cut growth vas observed was 2·4 kgf/cm.

Cut growth samples put on the repeated extension machine at ·alues of tearing energy less than 1·7 kgf/cm showed no cut growth ıfter repeatedly being stretched for 3×10^6 cycles. It was therefore ıssumed that the value of T_0 for the microporous polyurethane foam vas approximately 2 kgf/cm.

ı.2.4 Comparison with vulcanised rubbers

The majority of the published [23–25] cut growth test results which ıre expressed in terms of the parameter 'tearing energy' have been :onfined to vulcanised rubbers such as natural rubber (NR) and ;tyrene–butadiene rubber (SBR).

The cut growth results for the foam and solid polyurethanes are :ompared with NR and SBR data from published papers in Fig. 6.4.

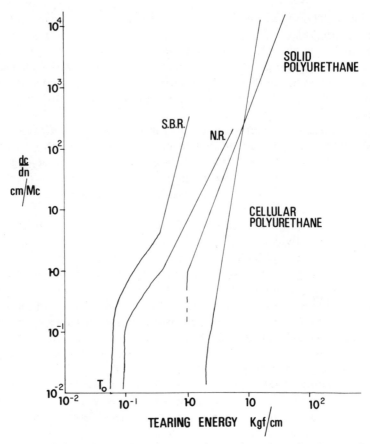

Fig. 6.4. Variation of rate of cut growth with tearing energy for foam and solid polyurethane compared with vulcanised NR and SBR elastomers.

The line drawn through the results for the solid polyurethane has slope on the double logarithmic scales of 2·5 compared to 6 for the foam polyurethane. The higher slope in the case of the foam polyurethane would be expected as no allowance has been made for the effect of the holes. The cut growth process in solid materials is also different as it is a continuous growth of a crack, whereas in the foam polyurethane it is a repeated process of crack initiation through the polyurethane solid strands and rapid growth through the holes.

The lower limiting value of tearing energy (T_0) at which no cu growth occurs for the solid material is approximately the same as found for the foam polyurethane. The effect of generating a cellular structure in a polyurethane material used in poromerics appears to have little effect on T_0 but increases the rate of cut growth with tearing energy.

Although the slope of the rate of cut growth–tearing energy curve is similar for polyurethane to NR, indicating as found in practice that both materials are highly hysteresial in character, the polyurethane is displaced along the tearing energy axis resulting in a far higher value for T_0 for polyurethanes than is normally obtained for vulcanised rubbers. This explains the good cut growth resistance of polyurethanes as found in practice and reasons for the high value of T_0 are given later in the paper.

6.2.5 Comparison with other shoe soling materials

The variation of cut growth per cycle with tearing energy for the random butadiene–styrene resin rubber (containing approximately 40 per cent of styrene) is shown in Fig. 6.5. These results are compared in the figure with a conventional pure gum vulcanised styrene–butadiene rubber (SBR) (23·5 per cent styrene) calculated in the same manner. It is clearly seen that the addition of a high styrene resin to the butadiene rubber increases the value of T_0. High styrene resin–butadiene rubber is known to be highly hysteresial in character. This effect is found in practice as microcellular resin–rubber soling materials are commonly manufactured from a mix consisting of styrene–butadiene copolymers with a high styrene resin content, and these have excellent resistance to cut growth in wear.

The variation of rate of cut growth with tearing energy for the styrene–butadiene block copolymers (thermoplastic rubber) is shown in Fig. 6.6. Investigations [27, 28] have shown that the structure of thermoplastic rubbers consists of long, flexible polybutadiene chains attached randomly to hard polystyrene blocks of approximately 300 Å in diameter and hence are highly hysteresial in character. It is interesting to note that the value of T_0 is high compared with conventional vulcanised rubbers and is similar in magnitude to polyurethane elastomers. The cut growth properties of the crosslinked thermoplastic rubber [29] are also shown in Fig. 6.6. It is seen that the introduction of crosslinks into the material and hence a reduction in its hysteresial properties considerably reduces the value of T_0. This is

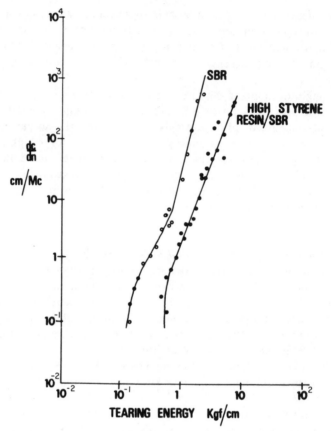

FIG. 6.5. Variation of rate of cut growth with tearing energy for random styrene
butadiene copolymer containing a high amount of styrene resin and conventional SBR
elastomer.

thought to be due to the crosslinking preventing the formation of the
typical thermoplastic rubber domain structure [27]. The non-forma
tion of this domain structure reduces the strength and hysteresis i
the material. This effect is similar to that found in polyurethan
elastomers, as shown in Fig. 6.7 where the cut growth properties of
linear cellular polyester polyurethane are compared with the result
from a crosslinked cellular polyether polyurethane of the sam
density. It is seen that the introduction of crosslinks, which in thi
case also prevent the ordered hard/soft segment domain structur

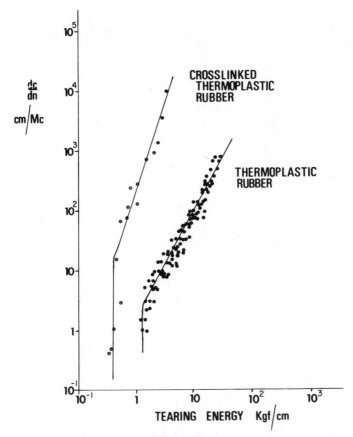

FIG. 6.6. Variation of rate of cut growth with tearing energy for thermoplastic rubber both uncrosslinked and crosslinked with 2·0 phr dicumyl peroxide.

being formed and reduce the strength and hysteresis of the material, considerably reduces the value of T_0.

6.2.6 Practical application of cut growth results

One of the main requirements for shoe soling and upper materials is a very high resistance to cut growth in wear and it is interesting to note that materials such as resin rubber, thermoplastic rubber and poly-urethane have a very high value of T_0, thus ensuring good cut growth resistance. Nevertheless, the demands of fashion are always with the footwear industry and flex cracking in wear of men's thick-wedge

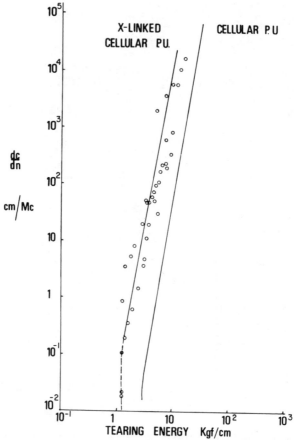

FIG. 6.7. Variation of rate of cut growth with tearing energy for linear cellular polyurethane elastomer compound compared to cellular crosslinked polyurethane.

flexible polyurethane-soled footwear is a current problem in the industry. During the early 1970s when polyurethane soling was first introduced, it was mainly used as thin units on conventional shoes and provided the compound was of a reasonable quality little cracking in wear occurred.

Similarly, the ladies' high-heeled platform footwear made in the fashion boom between 1972 and 1974 with polyurethane units caused little trouble in wear, the main reason for this being that the units were fairly rigid and tended to 'rock over' in wear rather than be lexed and thus cause high strains on the units which could lead to flex cracking failure.

However, in the last year there has been a trend towards the use of thick flexible wedge footwear in men's leisure footwear. Generally these shoes are of a fairly flexible upper and insole construction and hence flex a large amount, causing high strains on the sole and a tendency to flex cracking failure in wear as shown by the sole in Fig. 6.8.

6.3 SHOE ENGINEERING MEASUREMENTS

During recent years SATRA has undertaken a large amount of work [30] in the area of shoe engineering in order more clearly to define the stresses and strain on shoes in wear and therefore to determine the minimum physical properties required in the component parts of the shoe.

As part of this programme of work the surface strains on shoe soles in wear have been measured. This has been done by printing a grid on the bottom of the sole and photographing it through a glass plate in a walkway before and after flexing. Recently, strains on soles in wear have been measured by use of a sliding thread gauge [31]. Maximum strains measured on polyurethane sole units in wear in the flexing area are shown in Table 6.1. It is possible that local high strain concentrations in sole units in wear could be much higher than those quoted.

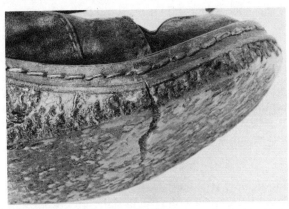

FIG. 6.8. Typical thick polyurethane wedge sole showing flex cracking failure.

TABLE 6.1
Maximum strains on polyurethane sole units in wear in the flexing area

Type of shoe	Maximum strain (%)
Thin, woman's unit	7
Woman's high-heeled platform unit	8
Thin, man's unit	11
Man's thick wedge	12
Man's thick wedge with wearing surface abraded	19
Man's thick wedge with cleated sole unit	over 30

Generally it has been found that flex cracking in wear is caused by cuts and cracks in sole units in wear, which can often measure 2 mm. Standard wear trials to detect liability for flex cracking are carried out by placing a 2-mm cut in soles and these are expected to last for a six-month wear period. These results from wear have been compared with the values obtained from cut growth theory.

6.3.1 Practical importance of lower limit of tearing energy (T_0)

Having established that the tearing energy theory of cut growth and fatigue failure is applicable to cellular polyurethanes [32] of the type used as shoe soling materials, it is necessary to consider the practical importance of the lower limit of tearing energy termed T_0.

If the material is stressed to strains in wear which correspond to tearing energy values lower than T_0, then inherent cracks in the material will not propagate. At values of tearing energy above T_0, however, the cracks can propagate and lead to failure such as flex cracking of the shoe soles shown in Fig. 6.8. It has been shown that T_0 is a basic strength property of a polymer and is a function of the intrinsic strength and flexibility of the molecular chain.

It was shown earlier that T_0 for a foam polyurethane used in upper materials was approximately $2 \cdot 0$ kgf/cm. T_0 for cellular polyurethane soling materials have been found to be approximately 1 kgf/cm.

The minimum tearing energy (T_0) is a function of strain energy density and initial flaw size (c_0) and hence from the general tearing energy equation can be expressed as

$$T_0 = 2KWc_0 \qquad (6.3)$$

Provided therefore that values of the right-hand side of this equation remain below 1 kgf/cm for soling materials, no cut growth will occur and the sole will not fail in wear. The value $2KW$ is, however, a

unction of strain and therefore as the strain increases, the critical
law size, to initiate failure of the sole, will decrease. This relationship
or a typical polyurethane soling material is shown in Fig. 6.9.

At strains less than 10 per cent fairly large flaws in the material can
be tolerated and will show no cut growth as the line in Fig. 6.9 is an
asymptote to the strain axis. This is observed in practice as flaws well
over 1·0 cm will not generally grow in the heel or waist of a shoe sole
since the strain on these parts of the sole in wear is generally less
than 10 per cent.

According to Fig. 6.9 it would appear that if a sole had a cut of
2 mm placed in it during wear then it would grow and cause flex
cracking of the sole if it were repeatedly strained by 25 per cent
during wear. From the strain measurements in wear shown in Table
6.1 it would appear that strain levels of this order can be quite easily
obtained on men's thick-wedge footwear, particularly if the units had
been worn for some time and the surface had been abraded off. The
strains would be much higher if the design incorporated some cleating
and hence would be most likely to fail in wear.

The cut growth results developed from the tearing energy theory
developed for vulcanised rubbers therefore appear to explain the flex

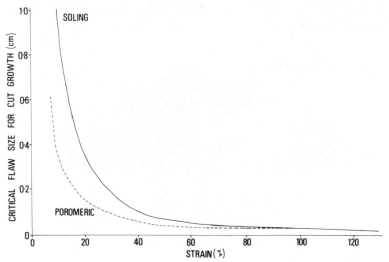

FIG. 6.9. Variation of cut length to produce failure with strain for cellular polyurethane
soling material and polyurethane foam poromeric material.

cracking failure of men's thick-wedge polyurethane-soled footwear in practice.

6.3.2 Application to foam polyurethane upper materials

The relationship between strain and length of cut to initiate cut growth in wear for a poromeric polyurethane foam upper material is shown as well in Fig. 6.9. Failure in wear with these upper materials can often occur from stitch holes if the material is not reinforced as shown by the wear return failure in Fig. 6.10.

A needle hole can impose a 1-mm cut in an upper material in wear and it is quite possible to strain the material over 30 per cent in wear, which would, from theory, lead to cut growth and failure in wear as shown (Fig. 6.10). It is for this reason that some form of reinforcement (e.g. nylon tape) is used in vulnerable areas in foam polyurethane shoe upper materials.

6.4 CONCLUSIONS

The theories of strength and reinforcement of vulcanised rubber developed over a number of years have been applied to give a satisfactory reason for the high strength properties of cellular polyurethane materials used as upper and soling materials in footwear. It has been found that the high strength, abrasion resistance and cut

FIG. 6.10. Failure from stitch holes in an unreinforced polyurethane foam poromeric material.

growth resistance is due to the incorporation of the hard urethane segments in the polyether or polyester rubber chain which act as minute, well-dispersed filler particles to produce a very effective 'self-reinforced elastomer'.

It has been found that the value of T_0 (the minimum tearing energy under which no cut growth takes place in the absence of chemical effects) is far higher for polyurethane rubbers than for vulcanised rubbers. This explains to some extent their good cut growth properties and the theory has been used to explain current failures of thick flexible wedge polyurethane units in men's footwear and cut growth from stitch holes in cellular polyurethane upper material.

Strains on soling materials in wear have been directly measured in shoe engineering studies and, knowing the likelihood of these materials to be cut by flints or glass in wear, it has been shown that the strains imposed are above the cut growth limit (T_0).

It is interesting to conclude that a combination of Bob Payne's two main scientific interests, design engineering of polymers and rubber physics theory, have been used to provide a satisfactory explanation in practical problems in industry, the latter being one of his main objectives throughout, regrettably, his too brief career.

ACKNOWLEDGEMENTS

The author is indebted to several useful discussions with SATRA staff during the course of this work and in particular to Mr P. J. Perkins for providing the shoe engineering measurements.

REFERENCES

1. J. A. C. HARWOOD, L. MULLINS and A. R. PAYNE. *J. Inst. Rubb. Ind.*, 1967, **1**, 17.
2. J. A. C. HARWOOD, A. R. PAYNE and R. E. WHITTAKER. (1969). In: *Proceedings of IRI Conference on Advances in Polymer Blends and Reinforcement*, Loughborough, UK, September 1969.
3. J. A. C. HARWOOD, A. R. PAYNE and R. E. WHITTAKER. *J. Appl. Poly. Sci.*, 1970, **14**, 2183.
4. J. A. C. HARWOOD, A. R. PAYNE and R. E. WHITTAKER. *J. Macromol. Sci.*, 1971, **B5**, 473.
5. A. R. PAYNE and R. E. WHITTAKER. *J. Inst. Rubb. Ind.*, 1970, **4**, 107.

6. R. E. WHITTAKER. *Poly. Age*, 1971, **2**, 21.
7. L. G. HOLE and R. E. WHITTAKER. *J. Mat. Sci.*, 1971, **6**, 1.
8. R. E. WHITTAKER. *J. Coated Fibrous Mat.*, 1972, **2**, 3.
9. A. N. GENT and A. G. THOMAS. *Rubb. Chem. Tech.*, 1963, **36**, 597.
10. R. E. WHITTAKER. *J. Appl. Poly. Sci.*, 1971, **15**, 1205.
11. H. OERTEL. *Textil-Praxis*, 1964, **19**, 820.
12. H. OERTEL. *Bayer Farbenrev.*, 1965, **11**, 1.
13. H. RINKE. *Angew. Chem.*, 1962, **74**, 612.
14. R. BONART. *Kolloid Z.*, 1966, **211**, 14.
15. R. BONART. *J. Macromol. Sci.*, 1968, **B2**, 115.
16. V. M. SHIMANSKII, S. I. SHKOLNIK and S. B. KOZAKOV. *Soviet Rubb. Tech.*, 1967, **26**, 20.
17. R. BONART, L. MORBITZER and G. HENTZE. *J. Macromol. Sci.-Phys.*, 1967, **B3**, 337.
18. S. B. CLOUGH and N. S. SCHNEIDER. *J. Macromol. Sci.-Phys.*, 1968, **B2**, 553.
19. S. B. CLOUGH, N. S. SCHNEIDER and A. O. KING. *J. Macromol. Sci.-Phys.*, 1968, **B2**, 641.
20. J. H. SAUNDERS and K. C. FRISCH. (1962). *Polyurethanes, Chemistry and Technology.* Interscience.
21. R. S. RIVLIN and A. G. THOMAS. *J. Poly. Sci.*, 1953, **10**, 291.
22. A. A. GRIFFITH. *Phil. Trans. Roy. Soc.* (*London*), 1920, **A221**, 163.
23. G. J. LAKE and P. B. LINDLEY. *J. Appl. Poly. Sci.*, 1964, **8**, 707.
24. G. J. LAKE and P. B. LINDLEY. *Rubb. J.*, 1964, **146**, 24.
25. A. N. GENT, P. B. LINDLEY and A. G. THOMAS. *J. Appl. Poly. Sci.*, 1964, **8**, 455.
26. H. W. GREENSMITH. *J. Appl. Poly. Sci.*, 1963, **7**, 993.
27. G. HOLDEN. *J. Elastoplast.*, 1970, **2**, 234.
28. E. T. BISHOP and S. DAVISON. *J. Poly. Sci.*, 1969, **C26**, 59.
29. C. M. BLOW and R. E. WHITTAKER. *J. Appl. Poly. Sci.*, 1974, **18**, 3443.
30. R. E. WHITTAKER. (July 1975). *Physics in Technology*, p. 150.
31. P. J. PERKINS. *SATRA Bulletin*, 1977, **17**, 502.
32. R. E. WHITTAKER. *Brit. Poly. J.*, 1972, **4**, 437.

Chapter 7

THE FATIGUE BEHAVIOUR OF THERMOPLASTIC RUBBER AND OTHER SOLINGS

D. PETTIT and N. A. MILLER

7.1 INTRODUCTION

There are four properties of shoe soling materials which can be said to be of major importance in wear: durability, grip, shape stability, and flex cracking resistance. The last of these is the one which is the most critical in that if it is inadequate and the sole cracks within the reasonable life of the shoe, it will cause the wearer sufficient concern to return the shoes, often justifiably, and thus incur expense to the shoe manufacturer. It is therefore, reasonable for the manufacturer to give more attention to this property of solings than to the others.

7.2 FAILURE PHENOMENA

The various types of premature failure are shown in Figs. 7.1–7.3. A man's relatively-thin (7-mm) PVC sole of hardness IRHD 68, shown in Fig. 7.1 has cracked in wear. The polyurethane sole in Fig. 7.2 has cracked through a combination of design features which lead to high strain in wear plus possibly below-par physical properties which occasionally occur with this relatively variable polymer. The thermoplastic rubber sole in Fig. 7.3 has cracked primarily because of bad design features causing excessive local strain.

The factors linking high surface strain with cut growth in wear, leading to cracking failures are examined in Chapter 6 by R. E. Whittaker who deals with the concept of minimum tear energy, T_0, with particular respect to urethane polymer soling. This approach

FIG. 7.1. Cracked plasticised PVC shoe sole.

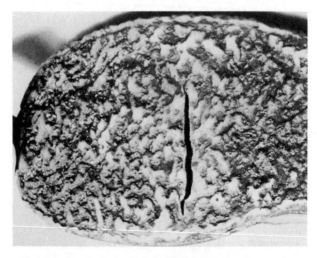

FIG. 7.2. Cracked reaction-moulded cellular urethane sole.

gives a clear understanding of the problem of sole cracking but involves a great deal of experimental testing.

7.3 TEST METHODS

In the practical situation such testing is usually either uneconomic or impossible, owing to lack of sufficient suitable samples. One is there-

FIG. 7.3. Cracked thermoplastic rubber (SBS) sole.

fore forced largely to rely on routine flexing instruments to give an empirical comparison with previous experience. A multitude of test equipment designs has been, and is, used for this testing and some of these will now be illustrated. Figure 7.4 shows the Hall flexer, suitable for relatively thin solings, and tests strips of plain soling material by a rolling action between two reciprocating parallel plates at a distance of 32 mm and a rate of 255 flexes per minute. Figure 7.5 shows the

FIG. 7.4. Hall flexing machine with sample clamped between reciprocating parallel plates.

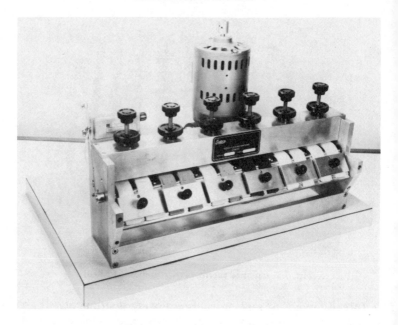

FIG. 7.5. Ross flexing machine.

Ross machine (modified) which flexes a strip of material, in which a standard 2-mm cut has been placed, through 90° over a 9-mm rod. The test strip is clamped at one end only and in the SATRA version is flexed 60 times per minute, usually at −5°C to offset heat build-up in the sample. Figure 7.6 shows a development by PFI in Pirmasens of the De Mattia test. In the PFI tester, the sample is clamped at both ends and flexed through 90°, 150 times per minute, over a 30-mm diameter roller. The sample is usually 10-mm thick, and backed with a 2-mm cellulose insole board which partially reproduces the flexing conditions in wear and certainly increases the surface strain on the flexed sample. This machine, like the next machine, the belt flexer, shown in Fig. 7.7, is very suitable for testing complete sole units and investigating the cracking tendency of sole surface patterns. In this test the sole units are stuck to a strip of canvas which is driven round a pulley of 56-mm diameter about 100 times per minute.

All these tests reproduce to a greater or lesser degree the flexing action of the shoe, and surface strain measurements have been carried out on samples flexed (statically) on the machines. Figure 7.8 shows the relationship found between thickness and surface strain on

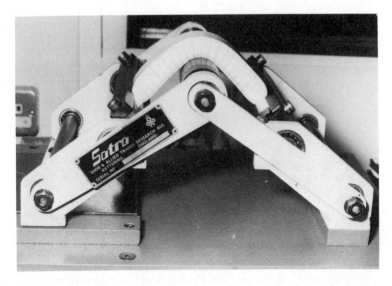

FIG. 7.6. PFI flexer with sample marked to show spread of surface extension beyond the flexed region.

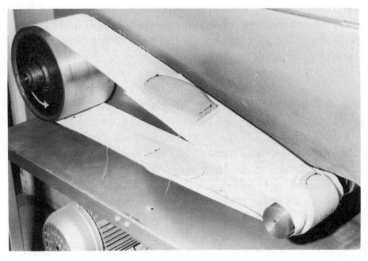

FIG. 7.7. Belt flexer used to establish the effect of sole surface patterns on cracking tendency.

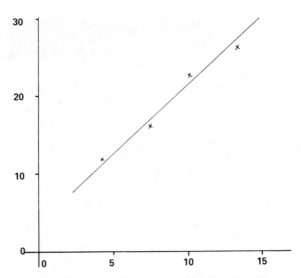

FIG. 7.8. The relationship between the thickness (abscissa, in millimetres) and observed surface strain (ordinate, per cent) of a urethane (skin-on) Ross fatigue specimen.

urethane soling strips having the moulded sole pattern still present. The strain is measured by non-return slippage of a wire clamped to the flexed surface beneath staples. The theoretical geometrical surface strain is roughly twice that shown.

When we use backed (by cellulose insole board) urethane samples on the Ross or the PFI flex tester, the relationship between thickness and surface strain is not regular. Strains of 30–37 per cent are achieved. The geometry of the PFI flex tester would lead one to expect surface strains of about 100 per cent but reference to the sample shown in position in Fig. 7.6 illustrates why the surface strain is so irregularly linked with thickness. The originally parallel lines drawn on the edge of the flexed specimen, normal to the flexed surface show that the strain over the flexed region is spread to adjacent non-flexed areas. Increase in thickness of soft cellular soling compounds will not only increase the surface strain but will also contribute to its easy relief by shear in the sample. The expected surface strain is also diminished, to a lesser extent, by self compression of such samples during flexing.

This really points out the dilemma facing the soling technologist; whether a material should be tested in as predictable and reproducible fashion as possible (*e.g.*, repeated linear extension, with or without cuts) or tested in a way which reproduces some of the wear conditions (*e.g.*, PFI flexing with an insole attached to the back). In either case, an accelerated test is required in order to provide answers in as short a time as possible. Although the tests are meant to predict cracking failures (frequently being accelerated by lowering the test temperature, by increasing the frequency of flexing deformation or the degree of extension compared with that met in wear), application of the WLF or time/temperature shift principle to our results has met with little success. The reason for this seems to be the build-up of temperature which occurs during testing. For instance, during the Ross testing of a PVC soling compound a sufficiently large temperature rise occurs to halve the modulus.

7.4 TESTS APPLIED TO THERMOPLASTIC RUBBER

The purpose of this paper is largely to present data based on repeated extension of thermoplastic rubber (compounded SBS block copolymer shoe soling) and to make comparisons with similar results for plasticised PVC. This comparison is particularly interesting because in practice the SBS shows superior flex crack resistance. It must be said that the SBS compounds* used as solings represent a wide range of formulations which typically include the base polymer, extending oils and resins such as crystal polystyrene in addition to mineral fillers. The particular composition affects behaviour of the polymer in processing and testing and therefore some of the following results must be interpreted with caution. Initial interest in the fatigue failure of SBS polymers occurred during an investigation of the beneficial effects of tyre tread crumb as an additive to SBS soling compounds. It was expected that the rubber crumb particles would act as crack stoppers and thus prolong the fatigue life of SBS. Figure 7.9 (from work by P. H. Kellett) shows a substantial increase in the number of extensions (at 90 per cent strain) to failure. It was necessary in this test to include a Banbury milled control (B in Fig. 7.9) as the mixing stage caused some deterioration in properties of the SBS.

Surface strain measurements on thick soles on flexible shoes have

---

*SBS compounds used were the Kraton TR materials of Shell Chemical Co.

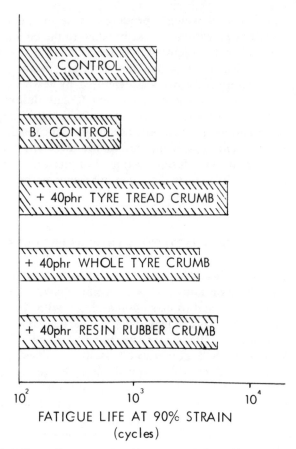

FIG. 7.9. The effect of type crumb rubber on the fatigue life of SBS thermoplastic rubber. Banbury milled (B) control has the same milling history as blended samples.

suggested that an overall strain at the flexed portion of a plain sole does not rise above 20 per cent in normal walking but it is believed that detrimental sole patterns could raise this by a factor of 2–3. In fact, a cracked SBS sole similar to that shown in Fig. 7.3 had a surface extension of 50 per cent in the troughs of the pattern when flexed in the PFI flexer.

The 90 per cent extension used for the fatigue experiment with the tyre crumb SBS compound is unlikely to be met in wear, but it is conceivable that 50 per cent surface extension could be achieved in wear with poor sole patterns. Furthermore, 50 per cent extension is about the lower limit of strain which can be produced with reasonable accuracy on the previously described cyclic extension test machines and hence was used for the following investigations.

Table 7.1 shows the number of extensions to failure of dumb-bell specimens of various SBS and PVC solings subjected to cyclic tests at room temperature with the number of extensions per minute being 120. The specimens in this constant extension device showed a certain amount of extension set, as would be expected, and it was therefore decided to repeat this experiment on the machine shown in Fig. 7.10. This was designed to take up the permanent set achieved in the test, and to ensure that the specimens were always under positive stress. The weight of the lower ratchet device was 1 kg, which was insufficient to cause detectable strain on the 5-mm square dumb-bells used. The machine (60 extensions per minute) indicated that the SBS benefitted from these conditions of continuous positive stress as is shown in Table 7.2, whereas the PVC gave a much lower number of cycles to failure. The constant extension machine was modified to carry out these comparative extensions at the same rate as the continuous stress machine (60 extensions per minute).

The behaviour of typical SBS and PVC soling compounds was also compared using the Ross test at −5°C and room temperature with results shown in Table 7.3. It is apparent from these results that the lower temperature is beneficial to SBS whereas PVC, like most other solings, gives a much poorer flex performance.

TABLE 7.1
Fatigue failure comparisons[a]

Material[b]	Hardness (IRHD)	Number of cycles to failure (kc)
PVC (DIM)	62	No failure up to 10^3 kc
PVC (unit)	75	No failure up to 10^3 kc
Thermoplastic rubber (SBS)	58	146 kc
Thermoplastic rubber (SBS)	85	95 kc

[a] Repeated extension at: room temperature; 120 cycles·min^{-1}; initial extension *ca.* 50 per cent.
[b] DIM, direct injection-moulded sole; unit, separately moulded sole attached to a shoe with adhesive.

FIG. 7.10. Continuous load repeated extension machine. Extension set in the samples is take up by the heavy (1-kg) ratchet device attached to the lower sample clamp.

Figure 7.11 shows the stress–strain curves for SBS and PVC at the two test temperatures. As one would expect from the T_g values of the two polymers, the PVC shows a substantially greater increase in modulus than does the SBS.

However, in shoe soles we are probably only concerned with the stress–strain curves of these two materials up to extensions of about

TABLE 7.2

Comparison of fatigue behaviour under conditions of either continuous load or constant extension at: room temperature; 60 cycles·min⁻¹; initial extension ca. 50 per cent.

| | | Number of cycles to failure (kc) | |
	Hardness (IRHD)	Continuous load (kc)	Constant extension (kc)
PVC (DIM)	62	316	≫750
PVC (unit)	75	203	≫750
Thermoplastic rubber	58	477	118

DIM, direct injection-moulded sole; unit, separately moulded sole attached to a shoe with adhesive.

TABLE 7.3

Gross cut growth behaviour of SBS and PVC at room temperature (RT) and −5°C

| | | Cut growth rate (mm·kc⁻¹) at | |
	Hardness (IRHD)	room temperature	−5°C
PVC (unit)ᵃ	75	nil	0·431
Thermoplastic rubber	55	0·170	0·010
Thermoplastic rubber	60	0·197	0·023
Thermoplastic rubber	86	0·116	0·030

Unit, separately moulded sole attached to a shoe with adhesive.

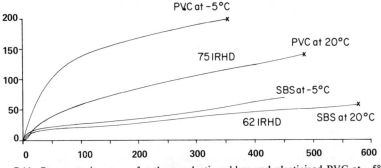

FIG. 7.11. Stress–strain curves for thermoplastic rubber and plasticised PVC at −5°C and 20°C. Ordinate: stress (kgf cm⁻²); abscissa: strain (per cent).

50 per cent and Figs. 7.12–7.15 show the effect of cycling the tw
polymers up to this level at room temperature and at −5°C. Thes
curves, which were prepared at 50 cm·min⁻¹ jaw separation spee
(about one-quarter of the rate at which strain takes place in walkin
or in Ross flexing), go some way to explaining the relative Ros
behaviour of these solings at −5°C and room temperature.

The SBS shows very significant softening after the first extension a
both temperatures. However, the area of the hysteresis loop at roon
temperature is tending to become very small with subsequent exten
sions compared with the area of the SBS loop at −5°C. This implie
that there is relatively little capacity for dissipating energy at th

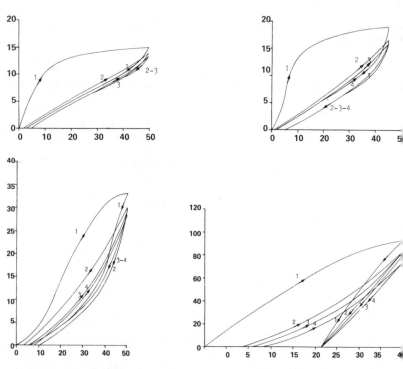

FIG. 7.12. Hysteresis of SBS at room temperature (top left).

FIG. 7.13. Hysteresis of SBS at −5°C (top right).

FIG. 7.14. Hysteresis of PVC at room temperature (bottom left).

FIG. 7.15. Hysteresis of PVC at −5°C (bottom right). In each case, stress (kgf cm⁻²) i
given on the ordinate and percentage of strain on the abscissa. Numerals on the curve
refer to the extension recovery cycle: *e.g.*, 1, first extension recovery cycle, etc.

:rack tip relative to the recoverable energy available to promote cut
;rowth at room temperature. The PVC, however, has a reasonable
·ysteresis area at room temperature. At −5°C the hysteresis loop at
·his speed cannot be interpreted because the speed of recovery of the
·ample is not great enough to follow the jaw speed.

The behaviour of SBS in wear is no doubt further complicated by
·the yield point which the stress–strain curves of some of the harder
·grades show. This is illustrated by Fig. 7.16 which shows variation in
·the yield point with speed of test. Walking speed is probably
·equivalent to the strain rate of about $25 \times 10^{-2}\,\mathrm{s}^{-1}$ in this figure.

It is quite likely with many types of sole that the surface exten-

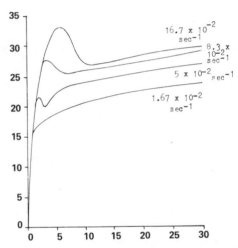

FIG. 7.16. Effect of strain rate on yield behaviour of thermoplastic rubber Ordinate:
stress (kgf cm^{-2}), abscissa: strain (per cent).

sion falls near the yield point of such an SBS. Recycling below the yield
point produces a negligible hysteresis loop whilst recycling above the
yield point gives rapid softening. Thus it seems likely that
some sole patterns could flex sufficiently to give soft SBS in very
limited areas surrounded by harder areas where the strain was
insufficient to lead to softening. This would undoubtedly lead to very
exaggerated strain concentrations but in practice during wear failures
are rare.

It is hoped that this paper will have helped identify the many
problems confronting the polymer physics approach as a practical

means of forecasting the fatigue behaviour of solings in wear. Such an approach may only be expected to be fruitful when comparing two materials which are sufficiently similar in properties to ensure that they will both experience the same degree of flex in wear on shoes. A future approach is intended to be the use of a limited bending moment test with made-up shoes and simulation of walking habits and shoe styles.

Chapter 8

RUBBER HYSTERESIS IN BIAXIAL AND TRIAXIAL LOADINGS

B. P. HOLOWNIA

8.1 INTRODUCTION

Hysteresis is a dissipation of mechanical energy as heat during deformation of the material. In a previous paper [1] the author investigated the relationship between hysteresis loss and filler loading. Addition of carbon black to a gum stock increases the hysteresis loss by an amount depending on the type and the percentage of carbon-black content in the rubber formulation. Because vulcanised rubber reinforced with carbon black exhibits enough hysteresis to make it one of the important factors in the selection of a compound for a given application such as tyres [2], much more attention has been paid to the phenomenon of hysteresis in recent years.

The simplified theory [3] shows that for a given rubber the percentage of hysteresis loss is dependent only on the loss angle δ, which is obtained from the elastic modulus E' and loss modulus E''. Hence most investigators, when studying hysteresis, have measured the loss modulus E'' or loss angle δ, directly in order to obtain the value of hysteresis loss.

Ulmer, Hess and Chirico [4] have shown that a considerable change in the loss angle δ can be achieved by using different types of carbon blacks. In another paper Ulmer, Chirico and Scott [5] show that the value of E'' changes significantly when a large variation of frequency occurs (0·01 to 100 Hz). The dependence of hysteresis on temperature [4], frequency variation [5], amplitude, and type of carbon black used [5–8] has been considered by many authors.

However, as far as is known, past work on the subject has been confined to uniaxial loading only where the values of δ, E'' or the out-of-phase shear modulus (G'') were measured [9] under specified frequency and deformation to obtain the hysteresis loss. In the present investigation the main purpose is to show that for a given rubber formulation, the amount of hysteresis loss is dependent on the stress distribution within the rubber component; hence a number of different rubbers were tested under uniaxial, biaxial and triaxial compression and the hysteresis loss was obtained for each type of loading.

To study the interaction of stress in different directions on the hysteresis loss, a complex stress system was set up using bonded rubber cylinders. The stress distribution within the cylinders was calculated on a computer by solving the classical equations of elasticity for each boundary condition, and the hysteresis loss obtained experimentally was compared with the stress distribution inside the rubber block.

8.1.1 Notation

D Diameter of rubber specimen.

E' In-phase elastic modulus.

E'' Out-of-phase loss modulus.

G'' Out-of-phase shear modulus.

H Hysteresis loss.

h Height of rubber specimen.

R Radius of a rubber block ($R = D/2$).

r, z, θ Cylindrical polar coordinates of radius, depth, and angle, respectively.

u, w Displacement components along r and z directions, respectively.

λ, μ Lamé constants.

ν Poisson's ratio.

σ_r Radial-stress component.

σ_z Longitudinal-stress component.

σ_θ Circumferential-stress component.

τ_{rz} Shear-stress component in r, z direction.

δ Loss angle ($\tan \delta = E''/E'$).

ϵ Half peak-to-peak dynamic strain.

Hypothetical assumptions necessary for the application of the dynamic-relaxation process are

   c Viscous damping factor.

$\dot{u}, \dot{w}$ First differential of u and w with respect to time.

  Δt Increment of time.

  ρ Mass density.

8.2 EXPERIMENTAL

8.2.1 Materials used

Conventional sulphur-cured natural rubber gum vulcanisates were prepared (Table 8.1) and cylindrical test specimens of 2·54-cm diameter and 5-cm height were cured in a mould. Each sample was then tested under uniaxial, biaxial, and triaxial compression. In addition, a number of 7·6-cm diameter cylindrical rubber blocks with the flat faces bonded to metal plates were prepared so that under uniaxial compression, a complex stress system was set up within the rubber.

TABLE 8.1
Composition of rubbers in parts by weight

Rubber	Component			
	N-90	N-60	N-48	B-70
Natural rubber (NR) RSS 1	100	100		
Natural rubber SMR 20			100	
Polysar 301 (butyl)				100
HAF carbon black	90	60	48	
SPF				70
Dutrex oil 729 UK	5	5		
Paraffin wax	2	2		
Zinc oxide	3	3	4·5	4·5
Stearic acid	2	2	2	1
Flectol-H	1·5	1·5		
Flexon 845 (mineral oil)				30
CBS	0·75	0·75		
Sulphur	0·5	0·5	2·5	1
TMT	0·75	0·75		1·3
MBTS				1
2-Morpholinothiobenzthiazole[a]			0·8	
Vrtard			0·5	
LE			1·5	

[a] Santocure MOR.

114   B. P. Holownia

To ensure the repeatability of the results four specimens, 2·54-c
high, of each mix were tested in a uniaxial compression on an Instr
machine. The hysteresis effect of each specimen was measured aft
20 repeated cycles at a very slow rate of frequency (approximate
0·03 Hz). The area enclosed by the two curves in Fig. 8.1 w
measured with a planimeter and expressed as a percentage of the tot
area, *i.e.*, the total strain energy. An average area of four identic
specimens was finally used in the results.

8.2.2 Triaxial compression
A triaxial compression has been achieved in an apparatus similar
that used to measure Poisson's ratio [10]. A schematic diagram
shown in Fig. 8.2. It consists of a thick mild steel cylinder in
which a rubber specimen is placed. The radial clearance betwee
the cylinder and a specimen is such that when an axial strain
approximately 8 per cent is applied, the specimen will fill the cylind
completely. The load is applied by a closely fitted piston. Th
load-deflection curve was obtained from the Instron machine direct
and was checked using a clock gauge.

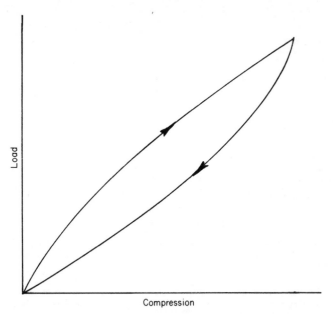

FIG. 8.1. Hysteresis effect of rubber under compression.

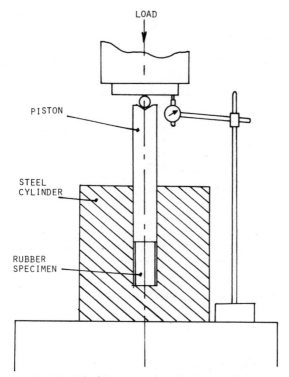

FIG. 8.2. Triaxial compression of rubber cylinders.

To determine the bulk compression of the rubber, an allowance is made [10] for the compression of the piston and the expansion of the mild-steel cylinder. Since the Poisson's ratio of the rubber is ·4998 the resulting stress distribution within the rubber specimen must result in a hydrostatic compression, which was checked theoretically using the same computer programme used on bonded rubber blocks.

The same specimens were subjected to the triaxial compression cycle varying between 0 and 8 per cent strain at a low frequency rate, and the hysteresis effect was measured after 20 repeated cycles using a planimeter in the same way as for the uniaxial compression.

8.2.3 Biaxial compression

A schematic diagram of a biaxial compression tester is shown in Fig. 8.3. It consists of a cylindrical perspex cover (A) fitted between two

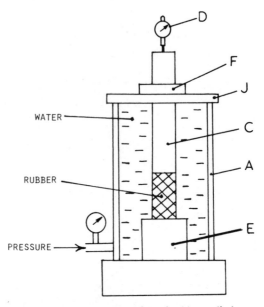

Fig. 8.3. Biaxial compression of rubber cylinders.

circular end-plates (J). The steel piston (C) was accurately ground and lubricated to minimise the friction between the piston and the bush (F) in which it ran.

The whole system was made watertight such that when it was pressurised to 8 atm, only a small seepage of water occurred between the free-moving piston (C) and bush (F). The biaxial compression was achieved by filling the perspex cylinder with de-aerated water and raising the pressure in the system.

To achieve biaxial compression a rubber specimen was glued to the base (E) and the piston (C) so that no load was applied to the end of the rubber cylinder. The load cycle consisted of pressurising the system detween 0 and 8 atm while deflection of the rubber specimen was measured with a clock gauge (D), thus obtaining a graph similar to Fig. 8.1 from which the hysteresis loss was determined using a planimeter as before.

Using this method, the pressure and the deflection of the rubber specimen were determined very accurately. The glueing, however

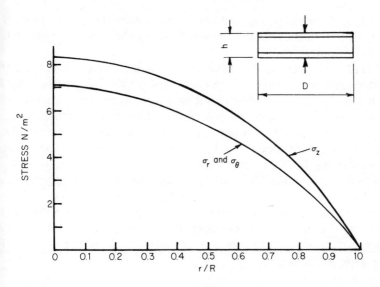

FIG. 8.4a. Stress variation within the rubber cylinder along the mid-plane under 10 per cent compression for N-90 rubber. $D/h = 6$; $\nu = 0.499$.

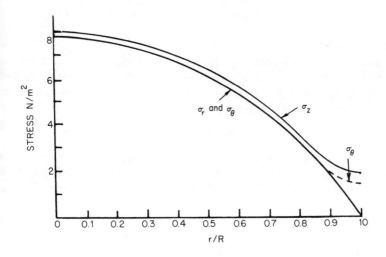

FIG. 8.4b. Stress variation within the rubber cylinder at the bonded face under 10 per cent compression for N-90 rubber. $D/h = 6$; $\nu = 0.499$.

resulted in shear stresses set up at the bonded faces and therefor
true biaxial stress was not obtained across the whole length of th
rubber cylinder.

Finally, to examine the effect of a complex stress system on th
hysteresis loss, a number of 7·6-cm diameter cylindrical rubber block
with the flat faces bonded to metal plates were tested under axi
compression. The strains were limited to 10 per cent so that classic
elastic theory could be used for determining the stresses inside
rubber block. The study was limited to blocks with diameter: heigh
ratios varying from 2 to 20.

Typical stress distributions in a bonded rubber block are shown
Fig. 8.4a and b, for $D/h = 6$, and Fig. 8.5a and b, for $D/h = 12$. The
figures show that as the height of the block reduces, the difference
between the stresses in the three directions becomes smaller, i.
there is a tendency towards a hydrostatic stress distribution as th
D/h ratio increases.

8.3 THEORY

For a visco-elastic material such as rubber, the energy loss per un
volume of material during one sinusoidal cycle is given by [3]

$$H = \pi \epsilon^2 E''$$ (8.

where ϵ is one half peak-to-peak dynamic strain and E'' is the visco
modulus. The above expression assumes an elliptical hysteresis loo

From eqn (8.1) a non-dimensional loss (H) can be obtained as
percentage of the total strain energy as

$$H(\%) = \pi/2 \cdot \sin \delta$$ (8.

where δ is a loss angle; hence eqn (8.2) shows that percentage
hysteresis loss is only dependent on the loss angle δ, which for
given rubber varies with temperature and can also vary with ar
plitude, and when a large change of frequency occurs.

Most investigators have used (8.1) to obtain H by measuring th
values of E'' directly under specified conditions. In the present wor
the amplitude was kept within 10 per cent strain and a very low ra
of frequency was used for all tests so that the change in the hysteres
was only due to the difference in the stress system within the rubb
specimen.

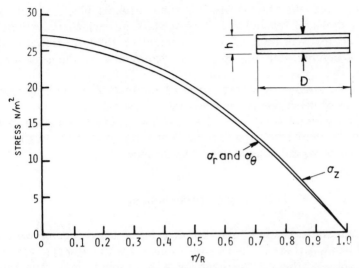

FIG. 8.5a. Stress variation within the rubber cylinder along the mid-plane under 10 per cent compression for N-90 rubber. $D/h = 12$; $\nu = 0\cdot499$.

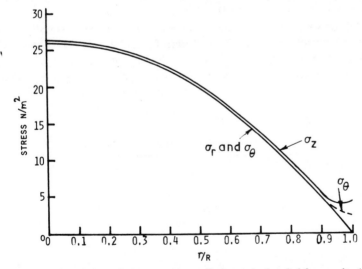

FIG. 8.5b. Stress variation within the rubber cylinder at the bonded face under 10 per cent compression for N-90 rubber. $D/h = 12$; $\nu = 0\cdot499$.

The stress distribution inside a bonded rubber block was obtained using dynamic relaxation [11] (Appendix 8.1) which is an iterative method that can be used on a computer to obtain numerical values which comply with the conditions determined by the equations of elasticity.

The accuracy and convergence of the method is dependent on number of mesh points and other factors [12, 13].

In the present study ν was taken as $0 \cdot 499$ and a fine mesh was used in the numerical computation to determine the stress distribution inside the triaxial, biaxial and bonded rubber blocks.

8.4　DISCUSSION

The theoretical results showed that for a triaxial compression, hydrostatic stress distribution was obtained within $0 \cdot 1$ per cent variation between the stresses over the whole rubber block. The experimental values (Table 8.2) of hysteresis under triaxial compression for all rubbers tested varied between 5 and $6 \cdot 5$ per cent. It was felt that a true triaxial compression could be achieved, the value of H would be even smaller, since in the present set-up the small movement of the rubber and the piston would inevitably have friction present which would contribute to the hysteresis loss.

A surprising result was that from the biaxial tests, a slightly greater value of hysteresis was obtained than from the uniaxial compression. This can be explained however by the friction in the piston of the apparatus used and also by the fact that almost the same change of shape of a rubber cylinder occurs when subjected to uniaxial or

TABLE 8.2
Hysteresis loss for rubber cylinders

Type of rubber	Hysteresis (% under 10% strain)			Young's modulus, E (MN/m^2)	Poisson's ratio
	uniaxial	biaxial	triaxial		
N-90	31·5	35·3	6·2	8·90	0·4994
N-60	21	24·5	5·5	4·75	0·4997
N-48	13·7	20·2	6·5	4·35	0·4998
B-70	22·8	25·3	5·4	2·93	0·4997

iaxial compression, which suggests that the hysteresis effect is
ependent on the change of shape of rubber block rather than the
tress level within the rubber. The theoretical results of biaxial
ompression have shown a large variation of stress distribution within
he rubber, particularly along the bonded face. Although the amount
f friction present was difficult to measure, the hysteresis losses
btained from the areas of four identical specimens were within 4 per
ent of each other.

Only two types of rubber, N-90 and N-60, were used to obtain the
alues of hysteresis loss variation with D/h. Figure 8.6 shows clearly
hat the hysteresis loss reduces significantly with increasing the D/h
atio. This is a very interesting result since there are numerous
veryday applications where the thin bonded rubber blocks are used.

8.5 CONCLUSION

he hysteresis loss in equitriaxial compression was very small (in the
rder of 6 per cent) for all rubbers tested, while for uniaxial and
iaxial compression, the values obtained were much greater.
xperimental results suggest that the hysteresis loss is dependent on

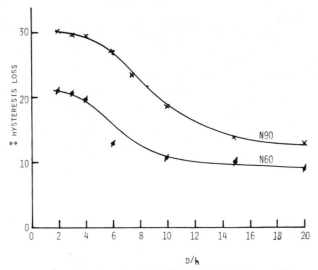

FIG. 8.6. Hysteresis loss against D/h for bonded rubber cylinders.

the change of shape of the rubber rather than the level of stresses present. In conclusion, it can be said that the results of the present work show that when hysteresis is an important factor, the type of stress system a rubber experiences can have a greater effect on hysteresis loss than does the rubber compound used.

ACKNOWLEDGEMENTS

The work described was started at the University of Natal, Durban, while the author was there on a sabbatical leave. The author wishes to thank the University of Natal, Durban, for the workshop assistance and their testing facilities, Dunlop Co. of Durban for producing some of the rubber specimens, and the Institute of Polymer Technology at Loughborough University for their help in preparing rubber specimens and for testing facilities.

REFERENCES

1. B. P. HOLOWNIA. 'Temperature buildup in bonded rubber blocks due to hysteresis'. *Rubb. Chem. Tech.*, 1977, **50**, 186.
2. P. R. WILLETT. 'Hysteretic losses in rolling tyres'. *Rubb. Chem. Tech.*, 1973, **46**, 425.
3. A. G. BUSWELL, E. GEE and E. R. THORNLEY. 'Dynamic testing and interpretation of results'. *J. Inst. Rubb. Ind.*, 1967, **1**, 43.
4. J. D. ULMER, W. M. HESS and V. E. CHIRICO. 'The effects of carbon black on rubber hysteresis. *Rubb. Chem. Tech.*, 1974, **47**, 729.
5. J. D. ULMER, V. E. CHIRICO and C. E. SCOTT. 'The effect of carbon black type on the dynamic properties of natural rubber. *Rubb. Chem. Tech.*, 1973, **46**, 897.
6. J. D. ULMER, V. E. CHIRICO and E. S. DIZON. 'Application of an equivalent strain of equal stress principle to the dynamic hysteresis of carbon-black-loaded and oil extended SBR'. *Rubb. Chem. Tech.*, 1975, **48**, 592.
7. A. R. PAYNE, R. E. WHITTAKER and J. F. SMITH. 'Effect of vulcanisation on the low-strain dynamic properties of filled rubbers'. *J. Appl. Poly. Sci.*, 1972, **16**, 1191.
8. A. S. MEDALIA. 'Elastic modulus of vulcanizates as related to carbon black structure'. *Rubb. Chem. Tech.*, 1973, **46**, 877.
9. A. R. PAYNE and R. E. WHITTAKER. 'Low strain dynamic properties of filled rubbers'. *Rubb. Chem. Tech.*, 1971, **44**, 440.
10. B. P. HOLOWNIA. 'Effect of carbon black on the elastic constants of elastomers'. *J. Inst. Rubb. Ind.*, 1974 (August), **8**, 4.

1. B. P. HOLOWNIA. 'Effect of Poisson's ratio on bonded rubber blocks'. *J. Strain Anal.*, 1972, **7**, 236.
2. J. R. H. OTTER. 'Computations of prestressed concrete pressure vessels using dynamic relaxation'. *Nucl. Struct. Eng.*, 1965, **1**(1), 61.
3. J. R. H. OTTER, A. C. CASSELL and R. E. HOBBS. 'Dynamic relaxation'. *Proc. Instn Civ. Eng.*, 1966, **35**, 633.

Appendix 8.1

In the axially symmetric case, as in the problem of circular rubber blocks [11] loaded axially considered here, only one of the three shear stresses is present and the circumferential stress is constant along any given circumference. Hence with reference to Fig. 8.7 the static equations are

$$
\left.
\begin{aligned}
\sigma_r &= (\lambda + 2\mu)\frac{\partial u}{\partial r} + \lambda\frac{u}{r} + \lambda\frac{\partial w}{\partial z} \\[2mm]
\sigma_\theta &= \lambda\frac{\partial u}{\partial r} + (\lambda + 2\mu)\frac{u}{r} + \lambda\frac{\partial w}{\partial z} \\[2mm]
\sigma_z &= \lambda\frac{\partial u}{\partial r} + \lambda\frac{u}{r} + (\lambda + 2\mu)\frac{\partial w}{\partial z} \\[2mm]
\tau_{rz} &= \mu\left(\frac{\partial w}{\partial r} + \frac{\partial u}{\partial z}\right)
\end{aligned}
\right\}
\tag{8.3}
$$

$$
\left.
\begin{aligned}
\frac{\partial \sigma_r}{\partial r} + \frac{\partial \tau_{rz}}{\partial z} + \frac{\sigma_r - \sigma_\theta}{r} &= 0 \\[2mm]
\frac{\partial \sigma_z}{\partial z} + \frac{\partial \tau_{rz}}{\partial r} + \frac{\tau_{rz}}{r} &= 0
\end{aligned}
\right\}
\tag{8.4}
$$

Equations (8.3) give the stress component in the three directions and eqns (8.4) state the conditions of equilibrium.

Equations (8.3) and (8.4) are transformed for the purpose of the calculations into the dynamic form by differentiating the stress–strain eqns (8.3) with respect to time and adding the inertia terms in eqns (8.4) as in the manner set out by Otter [12].

A damping term is added to the equilibrium eqns (8.2) and the full set of dynamic equations then become

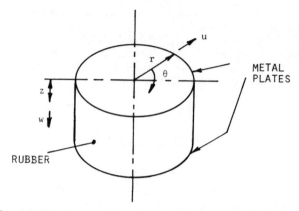

FIG. 8.7. Notation used in the stress–strain equations of elasticity.

$$\frac{\partial \sigma_r}{\partial t} = (\lambda + 2\mu)\frac{\partial \dot{u}}{\partial r} + \lambda \frac{\dot{u}}{r} + \lambda \frac{\partial \dot{w}}{\partial z}$$

$$\frac{\partial \sigma_\theta}{\partial t} = \lambda \frac{\partial \dot{u}}{\partial r} + (\lambda + 2\mu)\frac{\dot{u}}{r} + \lambda \frac{\partial \dot{w}}{\partial z}$$

$$\frac{\partial \sigma_z}{\partial t} = \lambda \frac{\partial \dot{u}}{\partial r} + \lambda \frac{\dot{u}}{r} + (\lambda + 2\mu)\frac{\partial \dot{w}}{\partial z}$$ (8.5)

$$\frac{\partial \tau_{rz}}{\partial t} = \mu \left(\frac{\partial \dot{w}}{\partial r} + \frac{\partial \dot{u}}{\partial z} \right)$$

$$\frac{\partial \dot{u}}{\partial t} + \frac{c}{\Delta t}\dot{u} = \frac{1}{\rho}\left(\frac{\partial \sigma_r}{\partial r} + \frac{\partial \tau_{rz}}{\partial z} + \frac{\sigma_r - \sigma_\theta}{r} \right)$$

$$\frac{\partial \dot{w}}{\partial t} + \frac{c}{\Delta t}\dot{w} = \frac{1}{\rho}\left(\frac{\partial \sigma_z}{\partial z} + \frac{\partial \tau_{rz}}{\partial r} + \frac{\tau_{rz}}{r} \right)$$ (8.6)

Equations (8.5) and (8.6) are then employed in the iterative method [12, 13] to obtain the values that satisfy them and meet the boundary conditions. The boundary conditions in the present problems are given in Fig. 8.8. The Lamé constants can be written in terms of D and Young's modulus, E

$$\lambda = \frac{E}{(1 + \nu)(1 - 2\nu)}$$ (8.7)

$$\mu = \frac{E}{2(1 + \nu)}$$ (8.8)

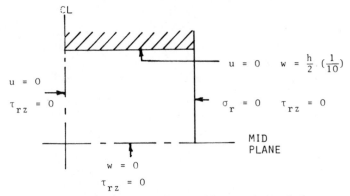

FIG. 8.8. Boundary conditions used in numerical analysis.

The convergence of the dynamic relaxation is dependent on the Poisson's ratio due to terms of order $1/(1-2\nu)$. Computer time for a solution of a given block increases rapidly as ν approaches 0.5. In the present investigation, ν was taken to be 0.499 which gives good results as was shown by Holownia [11].

Chapter 9

HAPE FACTORS FOR RUBBERS IN SHEAR AND COMPRESSION

P. G. HOWGATE

9.1 INTRODUCTION

All the data given in this paper were collected using the servo-hydraulic dynamic test equipment of RAPRA. A jig, which is sketch-outlined in Fig. 9.1, capable of applying compression to a simple shear experiment, was used throughout. Shown in Fig. 9.1 are the threaded bolts which are used to apply compression normal to the direction of shear over the range of compression from zero to approximately 80 per cent.

Shear deflection is measured in all cases as a relative displacement to the zero shear-force position with no applied compression. Shear forces are measured using a transducer with a very high lateral rejection ratio and a very high normal stiffness.

The upper frequency response of the measuring system was in excess of several hundred hertz, although most of the data presented in this paper consist of force–deflection characteristics which were measured on an X–Y recorder having a much lower frequency response. All force–deflection characteristics were obtained using a servo-hydraulic dynamic test machine in strain control; this enabled the instabilities obtained at high compressions and low shape factors to be studied in reasonable detail.

9.2 EXPERIMENTAL DATA

Figure 9.2 shows a graph of sine δ (loss angle) versus compression for three elastomer types. Observing this graph, sine δ shows a peak with all

FIG. 9.1. Jig capable of applying compression to a simple shear experiment.

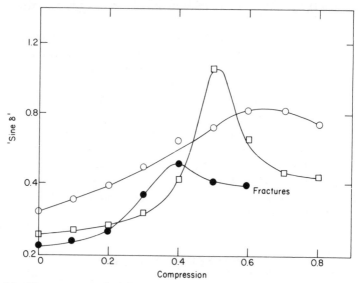

FIG. 9.2. 'Sine δ' versus compression for three types of elastomer. □, Natural rubber; ○, butyl rubber; ●, EPDM. Dynamic shear = ±0·5; temperature = 20°C; frequency = 0·3 Hz; static shear = 0; $h/d = 5/4$.

ree elastomer types, and particularly for the natural rubber sample.
 At this point it is worth looking more closely at the reasons for the
se of compression and shear combined deformations in engineering
ractice. The idea of applying a static compression is to soften the
ear elastic modulus to such an extent that, providing the specimen
 still undergoing the same order of deformation (and hence the
elastic component remains of the same order) the damping ratio will
crease markedly. Figure 9.2 shows this to be the case for all three
olymer types. Some interesting anomalies arise from Fig. 9.2. Care-
l observation of the peak value of sine δ for the natural rubber
ows it to exceed a sine δ equal to that of unity. The reason for this
ghlights the analysis problems which were encountered with the
hole of this experimental set of data.
 Figure 9.3 shows four shear force–deflection characteristics for
atural rubber at various applied static compressions. The obser-
ations to be made are simply that the characteristics are severely
on-linear at certain compressions. If one defines sine δ as the sine of
 phase angle between force and deflection, then obviously it has little
eaning with non-linearities of this order. In the case of the data
resented in Fig. 9.2, sine δ was measured as a simple ratio of
nergies. In this case the energy ratio is numerically equal to the
nergy lost per cycle divided by the energy used to strain the
lastomer unit during the same cycle, multiplied by a constant factor
 as to bring it in line with the linear theory. The key to the

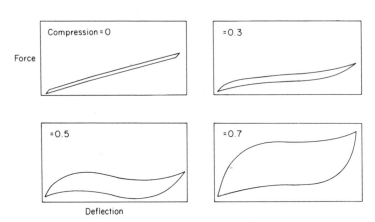

G. 9.3. Typical hysteresis curves for natural rubber at various static compressions.
Sinusoidal cyclic deformation. Shear strain $= \pm 1\cdot 0$; frequency $= 0\cdot 3$.

anomalous value of sine δ for natural rubber is this constant factor
On the linear theory, the force–deflection characteristic at maximum
damping, *i.e.*, sine $\delta = 1$, should become a circle. As can be seen from
Fig. 9.3 the force–deflection curves open more to that of a rectangle
than that of a circle. The net result of this is that more energy can
apparently be lost than is predictable on the linear theory. The
constant factor is used to allow for the difference in areas between a
rectangle and a circle and thus maximum possible loss on a thermo-
dynamic basis is equal to a sine δ value of 1·27.

The force–deflection characteristics in Fig. 9.3 also highlight the
problems of measuring a meaningful value of stiffness or modulus.
One cannot simply define a stiffness over the whole hysteresis curve.
Since both the value of measured modulus and the value of phase
angle are indeterminable, one cannot place any degree of reliability
upon any of the other extracted visco-elastic parameters. Bearing this
in mind this paper will present values of complex elastic and inelastic
stiffnesses and sine δ in order at this stage to show some of the trends.
The values are obtained adhering strictly to the letter of the linear theory
analysis, but inverted commas are used to remind the reader of possible
interpretation dangers.

The non-linearities of the combined deformations of shear and
compression affect not only the hysteresis curves but also most other
parameters involved. Figure 9.4 shows a set of hysteresis curves over
a range of dynamic shear deformations. There are two principal
points to note. First, at low shear deformations the hysteresis curve
reverts to linear behaviour and comparatively low damping ratio with
an associated high stiffness. As the dynamic deformation amplitude
increases, the non-linearities become more pronounced, the damping
ratio increases, and the stiffness falls. Secondly, the apparent centre
of the hysteresis curve, that is, the value of shear force at zero shear
deformation, is not in the geometric centre of the large-strain hysteresis curve.

The implications of the first factor are that the dynamic deformation amplitude has a considerable effect on the stiffness and
damping characteristics of the combination of deformations. Figure
9.5 and 9.6 highlight this effect. Figure 9.5 shows the variation of
'complex stiffness' versus dynamic shear amplitude. It can be seen
that the level of stiffness at low amplitudes approximates to that of
the simple shear modulus for the elastomer. As the amplitude increases, the stiffness decreases until above dynamic shear amplitudes
of 0·5 the value remains approximately constant.

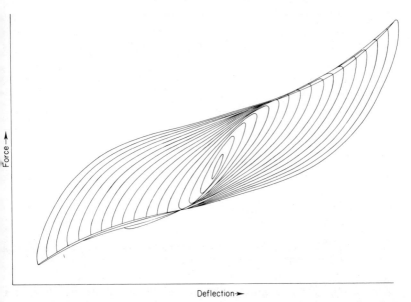

FIG. 9.4. Hysteresis curves of natural rubber at various dynamic shear strains. Compression strain = 0·4; room temperature; frequency = 0·3 Hz; max. dynamic shear strain ≃ 0·75; $h/d = 5/4$.

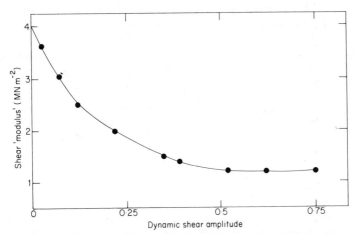

FIG. 9.5. Complex shear 'modulus' versus dynamic shear for natural rubber. Compression strain ≃ 0·4; temperature = 19·4°C; frequency = 0·3 Hz; static shear = 0; $h/d = 5/4$.

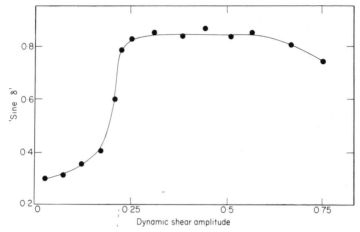

FIG. 9.6. 'Sine δ' versus dynamic shear amplitude for natural rubber. Compression strain $\simeq 0.4$; temperature = $19.4°C$; frequency = 0.3 Hz; static shear = 0; $h/d = 5/4$.

Figure 9.6 shows a graph for 'sine δ' versus dynamic shear amplitude, which at low strains approximates to the 'sine δ' value of the elastomer compound in simple shear alone. In this graph the transition between low and high 'sine δ' values is very sharp, occurring at a dynamic shear amplitude of approximately 0.2.

The trends of these characteristics show interesting behaviour when compared with the requirements of a combined engineering system, where energy absorption or shock damping needs to be combined with some low damping requirements such as anti-vibration.

The asymmetry of the zero shear-deformation force value leads us to a simple model for the exhibited behaviour. Figure 9.4 was constructed by first performing a large dynamic shear hysteresis loop and then stopping at the zero shear-deformation position after approaching from the right-hand half of the curve. The force value after a considerable time decayed to the value at the centre of the smallest deformation hysteresis loop. The subsequent curves were constructed by increasing the dynamic amplitude in stages and drawing each associated hysteresis curve on the same graph. If the zero shear-deformation point had been approached from the left-hand side of the curve, then the value to which the force would have decayed would have been displaced to the top of the curve instead of to the bottom. The new sequence of curves thus generated is simply the curves in

Fig. 9.4 rotated through 180° in the plane of the graph. The reason for
the duality of these zero shear-deformation-force values can be
simply explained by referring to experimental observation. With a test
sample of this height-to-diameter ratio, the sample will buckle under
this compression deformation. The sample can buckle in any direction
in the shear plane. Drawing a large-amplitude hysteresis curve before
allowing the system to relax in the zero shear-deformation position
ensures that the direction of buckling of the elastomer unit is always
in one direction. Since in the buckled condition, the resultant shear
force at zero shear deformation is finite and since the stored energy
of the unit opposes motion towards the zero force position, the
duality of force values arises naturally. Since the stiffness of the
hysteresis curve is low, this duality of force value and the hysteresis
which is associated with it tend to become significant compared with
any normal visco-elastic components. Figure 9.7 shows a curve which
is constructed by drawing a large-amplitude hysteresis curve similar
to that in Fig. 9.4 and stopping the cycle at various values of shear
deformation, allowing the force to relax for approximately 1 hr, and
then constructing a point. Once this has been performed for sufficient
points or at different shear-deformation values, an enclosed curve is
observed. This enclosed curve can be considered to be an additional
component of hysteresis owing to the duality of force values caused
by the specimen buckling. It now becomes apparent, that not only
does the reduction in elastic stiffness increase the damping ratio, but

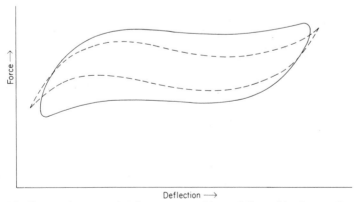

FIG. 9.7. Shear and compression frequency-sensitive and -insensitive hysteresis curves.
-----, Frequency-insensitive; ———, frequency-sensitive.

also there is a different inelastic component which is being generated by the combinations of deformations at certain shape factors.

In the use of a highly damped spring in an oscillatory mode the question which strikes the engineer is 'is there a heat build-up problem?'. In the case of a shear and compression system where the elastic modulus is simply softened, leaving the inelastic modulus unchanged, the heat build-up problem is no greater than the elastomer system in simple shear. The inelastic modulus tells the engineer in calories and joules and 'real-life' units how much energy will be dissipated per unit volume of elastomer. The additional 'inelastic' component of the true shear and compression system of low shape factor and its effect upon heat build-up can be deduced from Fig. 9.8. Accepting once again that the modulus figures presented in Fig. 9.8 have little meaning on the linear theory, they give some indication of trends relative to heat build-up. As can be seen from Fig. 9.8 the 'inelastic' component approximately doubles over a reasonable range of engineering compression strain. This will approximately double the heat build-up in the sample. The 'elastic' component drops over this range by nearly a factor of three. The gain in reduction of heat build-up compared with damping increase for the same spring stiffness (*i.e.*, more rubber) is therefore significant.

The errors in obtaining 'elastic' and 'inelastic' components by the linear theory are emphasised by Fig. 9.8 where at no value of compression does the elastic component become negative; however, Fig. 9.3 shows considerable negative-stiffness regions within the hysteresis curve which are applicable to the values in Fig. 9.8.

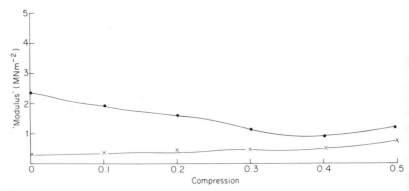

FIG. 9.8. Elastic and inelastic 'modulus' for natural rubber. Temperature = 19·2°C; shear strain = ±0·5; frequency = 0·3 Hz; $h/d = 5/4$. ×–×, G''; ●–●, G'.

With the likelihood of an increased level of energy dissipation, which will probably end up as heat, the question arises as to the temperature sensitivity of the achieved damping ratio. Figure 9.9 shows a graph of 'sine δ' versus temperature over a wide temperature range. Values of 'sine δ' are plotted at three compressions around the peak level of damping ratio. It can be seen that although at low temperatures 'sine δ' decreases, there is still a significant increase in damping ratio compared with the simple shear case until transitional damping takes over. This is contrasted with transitional damping itself, where the method of reduced variables tells us something of the temperature sensitivity of the mechanical characteristics over a temperature range of interest. If transitional damping is used to produce a damped spring then one suffers naturally from the high temperature sensitivity around the transition region. Compared with the temperature sensitivity well into the 'rubbery' region, this difference can be very significant to the engineer. For example, a high-nitrile rubber would have a transitional damping peak in approximately the right frequency spectrum at room temperature for a highly damped mechanical spring. If the useful damping peak is say 2 decades in frequency wide, then the method of reduced-variables data tells us that the transitional damping will disappear for a change in temperature of the order of 10°C. This compares favourably with the data in Fig. 9.9 where for a damping ratio greater than 0·6 at a compression of 50 per cent a change in temperature in excess of 60°C

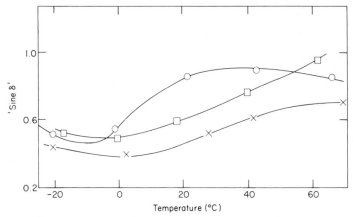

FIG. 9.9. 'Sine δ' versus temperature (natural rubber). Dynamic shear = ±0·5; frequency = 0·3 Hz; h/d = 5/4. Compression: ×–×, 4/10; O–O, 5/10; □–□, 6/10.

is needed to fall below that value. This temperature insensitivity highlights the effect of the use of geometry to control the dynamic characteristics. The data presented so far in this paper have been extreme to exemplify some of the highlighted properties. Certainly the naturally buckling specimen is not engineeringly useful in certain applications, where a negative stiffness region of the hysteresis curve is undesirable.

Figure 9.10 shows a graph of 'sine δ' versus reciprocal shape factor, ranging from a height-to-diameter ratio of 5:4 at one extreme to a height-to-diameter ratio of 1:4 at the other. It can be seen from Fig. 9.10 that there are still significant advantages in damping ratio to be achieved at shape factors which are considered highly stable. It should also be emphasised that the negative dynamic-stiffness characteristics achieved with some of the higher height-to-diameter ratio specimens can be removed by varying both the maximum dynamic shear and the maximum compression applied. The enclosed 'static' curve highlighted in Fig. 9.7 can also be measured for specimens having a higher shape factor than that demonstrated and can, in fact, be detected at a height-to-diameter ratio of 1:4, although very small. As stated earlier, this enclosed curve for high height-to-

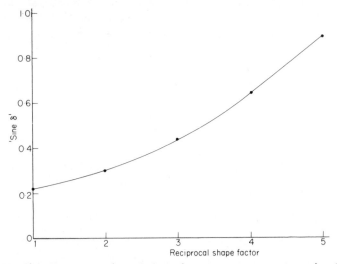

FIG. 9.10. 'Sine δ' versus reciprocal shape factor at constant compression (natural rubber). Compression strain = 0·5; static shear = 0; temperature = 19·4°C; frequency = 0·3 Hz; dynamic shear = ±0·5.

diameter ratio samples can be attributed by observation of the buck-ing phenomena. At low height-to-diameter ratios this is not the case. However, the duality of force at zero strain appears to be still present, and with a system unconstrained in the shear direction a duality of stable shear displacements will be observed. This obser-vation has been made within RAPRA during another research dealing with cylindrical elastomer bushes. The simple bush in compression is typical of a number of engineering bushes in that the elastomer is under a radial compression strain. At moderately high radial com-pression this is easily observed as a duality of positions in torsion which the bush can take on, depending on how it was assembled, and how it had previously been torsionally strained up to the point of observation. This fact is often ignored in design studies and has been known to lead to problems where the dynamic torque applied can change the torsional displacement of the bush from one state to the other.

The effects at low height-to-diameter ratios prompted RAPRA to look more closely at the model system and in particular the explana-tion based around the buckling phenomena. A current system being investigated is based around a potential energy model which can be simply derived for an idealised shear and compression case. The particular advantage of this model is that it looks upon the effect from the point of view of potential energy wells, rather than from the gross phenomenon of buckling itself. It thus has no restrictions based upon the geometry of buckling, merely upon the constraints which shape has upon the size of the energy wells. The particular advantage of this system to the engineer, is that most of the data can be presented in a graphical form which is easily assimilated from the point of view of design.

ACKNOWLEDGEMENTS

The author wishes to thank the Procurement Executive, Ministry of Defence for their funding and assistance in this work. Thanks are also due to Mr. D. P. Cottey for his assistance on this project.

Chapter 10

THE MEASUREMENT OF THE MECHANICAL RESPONSE OF VULCANISED NATURAL RUBBER TO SHEAR AND UNIAXIAL COMPRESSION

E. R. PRAULITIS, I. V. F. VINEY and D. C. WRIGHT

10.1 INTRODUCTION

When a constant stress is applied to rubber, the deformation is not constant but increases gradually with time; this behaviour is called 'creep'. Conversely, when rubber is subjected to a constant strain, a decrease of the stress in the material takes place; this behaviour is called 'stress relaxation'. Although creep and stress relaxation are allied phenomena, there is no simple general relation between them.

It is recognised that for quality-control purposes it has generally been sufficient to measure a single value of a given property under closely defined test conditions and the test procedures given in British and other national standards usually indicate only a 'single-point' value. On the other hand, satisfactory engineering design needs to be based on information covering both the magnitude and trends in the properties of materials under conditions simulating those experienced in practice.

The draft Standard (BS903: Part A15) describes the determination of creep in rubber and limits the methods in which rubber is deformed to tests in compression and in shear. It considers that measurements in tension are not relevant to product specifications since rubber is seldom deformed in this manner in practical usage.

The processes responsible for creep may be either physical or chemical in nature, and under normal conditions both processes will occur simultaneously. However, at ambient or low temperatures and/or short times, creep is dominated by physical processes whilst at high temperatures and/or long times, chemical processes are dominant. In general, physical creep is found to be directly proportional to logarithmic time, and chemical creep to linear time; hence it is neither safe to extrapolate time–creep curves in order to predict creep after periods considerably longer than those covered by the test, nor to use tests at higher temperatures as accelerated tests to give information on creep at lower temperatures.

10.2   APPARATUS

The generation of multipoint data by means of creep tests, to cover the range of loading times expected in service clearly requires some kind of automatic strain recording. Typically, this involves electronic transducers based on signals generated by resistive, capacitive, or inductive displacement elements. However, these are all analogue devices, which exhibit signal instability after long operating periods, resulting particularly from d.c. drift in both amplifier and recorder. Circuitry is available to improve the stability, accuracy and linearity but this may be expensive. A more suitable approach is to use a digital system, which is inherently immune to long-term drift, such as that afforded by Moiré fringe technology.

Although stability is an important requirement, high strain resolution is not. Long-term variation in specimen temperature cannot be controlled with confidence to better than $\pm0.5°C$. The high thermal expansion coefficient of rubbers—typically $>10^{-4}/°C$—would, in conjunction with this expected temperature variation, produce effects which would swamp the fine definition of a high-resolution instrument.

A Moiré fringe technique is used to detect the deformation. This system can be understood by reference to Figs. 10.1 and 10.2. A pair of linear diffraction gratings are arranged to produce a Moiré fringe image. The fringes move in sympathy with relative movement between gratings. A sinusoidally varying light contrast is thus produced at all points in the image. A grating displacement equivalent to one grating pitch induces a 360° phase change in light contrast at all points in the fringe image. Thus, if grating pitch is 0.0004 in., a pulse can be produced for each increment of 0.0001 in. of deformation. The signal generated by this varying light intensity is inherently digital and potentially free from d.c. drift induced instability.

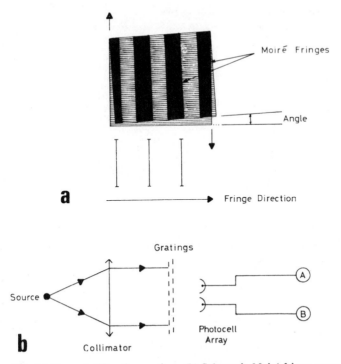

FIG. 10.1. (a) Crossed diffraction gratings. (b) Schematic Moiré fringe system.

The Moiré fringe image falls on a photodiode arrangement, consisting of four photodiodes connected in pairs to provide two electrical signals 90° out-of-phase. Each pair is connected in parallel opposition so as to obtain a signal which is insensitive to changes in ambient light level. Both signals are amplified and shaped into square wavetrains which are then fed into a gated logic circuit which discriminates the direction of each pulse. The time at which each forward or reverse strain increment occurs is then printed and the deformation totalised on a bi-directional counter.

Advantages of the Moiré fringe system include: (a) inexpensive automatic recording, (b) stability over long periods, and (c) strain is the independent variable.

Air bearings have been used to provide a frictionless active alignment system so that the load is applied along the specimen axis. The air bearings are accurately located in a platform which moves in the

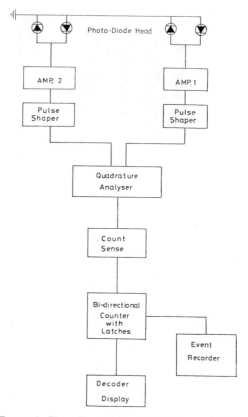

FIG. 10.2. Flow diagram of a Moiré fringe technique.

direction of the applied force. The bearing shafts are located in parallel rigid top and bottom plates.

For uniaxial compression, the compression platens are arranged above the bearing platform as shown in Fig. 10.3. The platens were ground parallel and chromium plated to provide a smooth, polished surface for testing lubricated samples. By using an 'O'-ring seal located in the bottom platen and a glass tube, swelling tests can also be carried out using this arrangement. Simple shear may be applied to a two-component sandwich arranged below the bearing platform as shown in Fig. 10.4.

The Moiré index grating is attached to the moving platform whilst the reference or 'scale' grating, light source and photodiode assembly

FIG. 10.3. Arrangement of compression platens for uniaxial compression.

are attached to either the top or bottom platen depending on the mode
of deformation. The use of air bearings thus provides the exten-
someter with complete freedom to move in the direction of strain,
complete rigidity against grating rotation and grating twist or movement normal to the grating plane. The gratings used are linear
chromium gratings with 2500 lines per inch. Fewer lines per inch
would, however, lessen the effect of temperature variations.

A 5:1 (25 cm:5 cm) lever arm, pivoted on a knife edge, amplifies
and transmits the load from the weight pan via chain and segment

FIG. 10.4. Simple shear applied to a two-component sandwich below the bearing platform.

couplings to the specimen. The use of flexible couplings ensures that lever-arm rotation does not affect the lever-arm ratio. Load application is controlled by a hydropneumatic system.

10.3 EXPERIMENTAL

The following examples of experimental work carried out using the testing machine have been chosen to illustrate its capability and versatility.

Compressive creep measurements have been carried out at 30°C and at near 50 per cent relative humidity. The testpieces used were discs, 13 mm in diameter and 6·3-mm thick, bonded to metal endplates of the same diameter with a cyanoacrylate adhesive.

Individual testpieces were deformed by a constant stress and the train determined as a function of time. The same specimens were then further conditioned by several short-duration deformations under the same stress followed by a 24-hr recovery period unstressed. Using the same applied stress, the strain was again determined as a function of time. The increase in strain with logarithmic time was found to be substantially linear in both cases.

The effect of cyclic loading and unloading on a filled natural rubber sample was examined by applying the compressive load for 200 s followed by a 20-s recovery unstressed, for a total of 10 cycles.

Using a step-loading procedure, compressive stress–strain curves have been constructed for unfilled natural rubber samples with shape factors (ratio of one loaded area to the total force-free area) ranging from 0·53 up to 2·89.

The ability of the apparatus to follow the swelling of rubber in a liquid was examined using 1-mm thick samples immersed in toluene. To allow free contact with the liquid, but still retain accurate location of the testpiece, two pieces of sintered glass were used with the sample sandwiched between them.

10.4   RESULTS AND DISCUSSION

A typical compression creep experiment, at a constant applied stress τ, is shown in Fig. 10.5. Over the time range given λ, the ratio of compressed thickness to the original thickness, is found to decrease in proportion with the logarithm of the time under load. Alternatively the compression strain $e = (1 - \lambda)$ increases with the logarithm of the time (t) under load. Figures 10.6 and 10.7 show how this slope, $de/d \log t$, varies with applied stress for vulcanised gum and ISAF carbon-black-filled natural rubber samples when the crosslinks are mainly monosulphidic (Fig. 10.6) and polysulphidic (Fig. 10.7). Even though comparison is made at constant stress, which to some extent compensates for a strain amplification [1], the effect of pre-stressing or conditioning is more pronounced for the filled materials. The alteration in slope brought about by pre-stressing at $2 \, \text{MN·m}^{-2}$ is illustrated in Fig. 10.8, for a range of filler concentrations, and the actual change, $\Delta(de/d \log t)$, produced shown in Fig. 10.9. The effect of pre-stressing is more noticeable the higher is the black concentration.

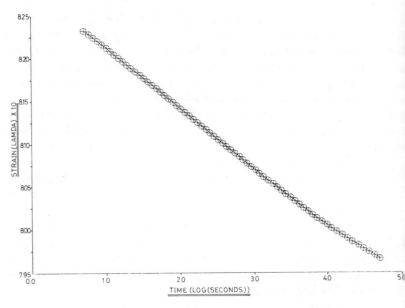

FIG. 10.5. Typical compression creep at a constant applied stress.

The rate of creep C, per cent per decade, may be defined as

$$C = \frac{1}{e}\left(\frac{de}{d \log t}\right) \times 100 \qquad (10.1$$

where e is the compression strain after 1 min. The changes in creep rate, at 20 per cent compression, brought about by the conditioning procedure, are shown in Figs. 10.10 and 10.11 for the same range of samples. Similar changes in stress relaxation rate have been reported by Derham [2] for filled natural rubber samples tested at 100 per cent extension. He attributed the changes brought about by pre-stressing to changes in intermolecular interactions different from those interactions which dominate in the glass transition region, indicating that a simple network theory of visco-elasticity is inadequate. Our results support the idea that the effects of pre-stressing are mainly topological changes in the network and not permanent cleavages of strong bonds.

The lowering of the creep rate by repeated stressing is confirmed by the results of the cyclic experiment shown in Fig. 10.12 and indicates that the process is 'complete' after 10 cycles.

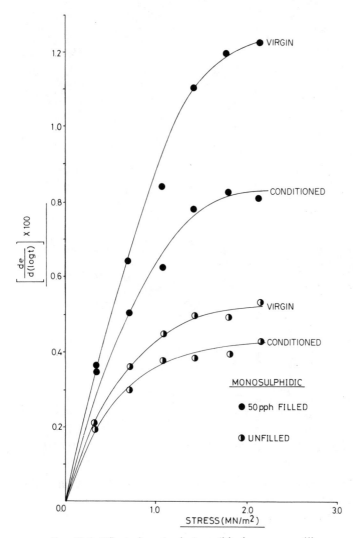

FIG. 10.6. Effect of mechanical conditioning on creep (1).

Clearly, tests must be carried out at strains appropriate to service conditions and it is important to ensure that the previous mechanical history of the specimens used in testing is appropriate to that of the actual components used in service.

Payne [3] suggested that the compressive stress–strain behaviour of

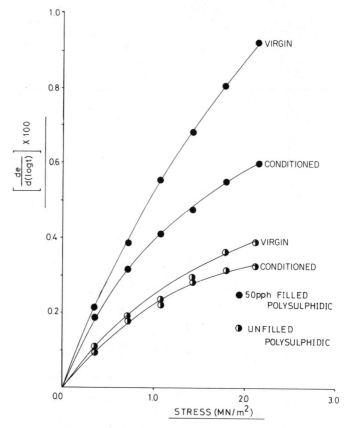

FIG. 10.7. Effect of mechanical conditioning on creep (2).

circular bonded-rubber pads can be represented by the relationship

$$\sigma = \frac{E}{3}(\lambda^{-2} - \lambda) \qquad (10.2)$$

where E is an apparent Young's modulus governing classically small compressions. This apparent Young's modulus is related to the shape factor S by [4]

$$E = E_0(1 + 2kS^2) \qquad (10.3)$$

where E_0 is the Young's modulus of the rubber and k an empirically determined factor less than the theoretical value of unity. Hence plots

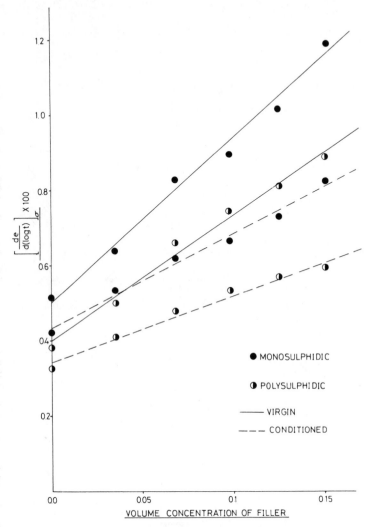

FIG. 10.8. Effect of filler concentration on creep.

of σ versus $(\lambda^{-2} - \lambda)$, for a given material, should be straight lines with gradients determined by S. The reason for the deviations shown in Fig. 10.13, which have also been observed in isochronous stress–strain curves, is not clear. Examination of other shape-factor theories, such as that proposed by Lindley [5], is now in progress and it is also hoped to ascertain the nature of the empirical constant k.

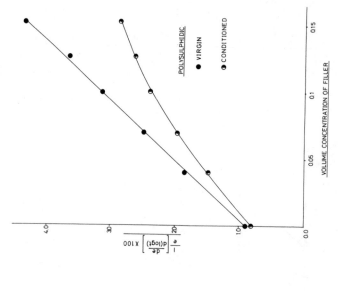

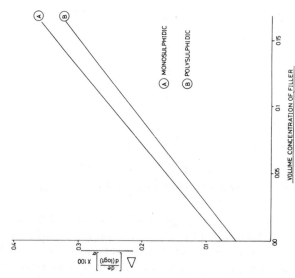

Fig. 10.9. Effect of filler concentration on the change in   Fig. 10.10. Effect of mechanical conditioning on

The results of the swelling tests are shown in Fig. 10.14 and give fractional volume swells for the two samples of 3·96 and 2·55. Similar samples tested using the RAPRA Hamden swell tester [6] gave 4·0 ± 0·1 and 2·6 ± 0·1, respectively. The values are seen to be in very good agreement.

According to Kraus [7] the reason for the swelling deficit is neither a catalytical influence of the black on the number of crosslinks during vulcanisation nor the formation of additional crosslinks on the filler surface, but adhesion of the elastomer to the carbon black particles.

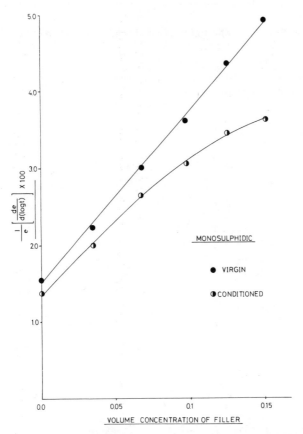

FIG. 10.11. Effect of mechanical conditioning on creep rate (2).

152　　　　　　　E. R. Praulitis, I. V. F. Viney and D. C. Wright

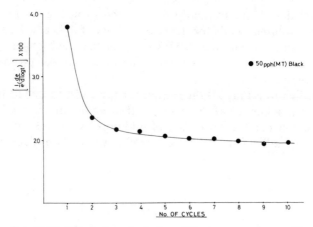

FIG. 10.12. Effect of mechanical conditioning on creep rate (3).

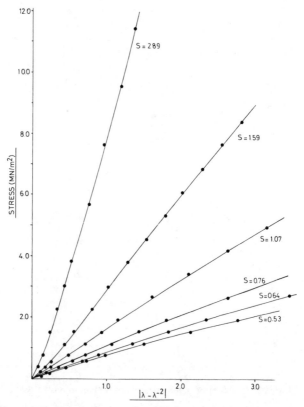

FIG. 10.13. Shape effects on bonded circular cross-section blocks.

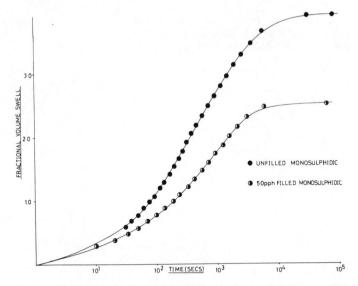

FIG. 10.14. Rate of volume swelling. Solvent, toluene; temperature, 25°C.

10.5   CONCLUSIONS

The machine described above by the authors forms a new device for the study of the deformation behaviour, in shear and compression, of rubbers under constant loads. It is capable of producing results to a high degree of accuracy, a capability which is being exploited in a number of investigations now in progress.

ACKNOWLEDGEMENT

The authors wish to thank the Science Research Council for financial support in the form of a CASE award for E. R. P.

REFERENCES

1. L. MULLINS and N. R. TOBIN. *J. Appl. Poly. Sci.*, 1965, **9**, 2993.
2. C. J. DERHAM. *J. Mat. Sci.*, 1973, **8**, 1023.
3. A. R. PAYNE. *Nature*, 1956, **177**, 1174.
4. A. N. GENT and P. B. LINDLEY. *Proc. Instn Mech. Eng.*, 1959, **171**, 111.

5. P. B. LINDLEY. *J. Strain Anal.*, 1966, **1**, 190.
6. R. P. BROWN. *RAPRA Research Report*, 1970, No. 191.
7. G. KRAUS. *Adv. Poly. Sci.*, 1971, **8**, 155.

Chapter 11

THE USE OF THE DIGITAL COMPUTER IN THE DESIGN OF RUBBER-BONDED-TO-METAL COMPONENTS

A. J. REED and J. THORPE

11.1 INTRODUCTION

For many years the designers of rubber-bonded-to-metal parts have used simple formulae to calculate some of the required design factors. These factors can be separated into seven commonly occurring sections as follows: (i) static deflection under load (stiffness), (ii) rubber stress (and strain), (iii) rubber damping factor, (iv) natural frequency, (v) dynamic amplitude, (vi) metal part stress, and (vii) heat generation within the rubber. These may have to be considered separately for both the individual bonded unit and the whole system in which the unit may be fitted in multiples or with other parts. Connected and peripheral factors to be considered are fatigue, creep, shrinkage on moulding, effects of environmental temperature and ageing.

Of the seven main factors, picked out because of their frequency of occurrence and immediate application nature, the simple formula technique will provide accurate design estimates for most applications of compression and for shear of simple rubber sections with respect to static deflection and average stress. Damping factors are well documented and correlate reasonably well with viscous approximation within the field under discussion. Dynamic characteristics are consequently easily handled for single degree of freedom applications, but problems in this dynamic situation are more usually related to the system involving a number of parts. Metal part stress cannot be simply handled on many occasions, with transmitted loading occurring through the rubber, whilst temperature estimation has been only crudely managed.

155

Many references and material relating to the basic design formulae available are contained in Lindley's booklet [1]. Extensions to these formulae have been under continuous development, particularly with regard to compression of bonded slabs and cylindrical 'bushes' Notable work on this subject has been carried out by Payne [2], Gent [3, 4], Lindley [1, 3] and Holownia [5]. There is a wealth of literature in this field, extending to high-strain analysis using finite elements [6] and to dynamic relaxation [5].

Bonded units, however, can usually be assessed in the low average strain region, where most are used. In this field, much work has been accomplished by the use of digital computers, particularly to analyse systems of bonded parts. Significant advances have been made in extending analysis techniques from the single degree of freedom systems to multiple spring/mass systems for vehicle suspensions and resiliently supported power units. The most commonly used computer methods produce natural frequencies, mode shapes plus static and dynamic deflections for such systems, assuming constant stiffnesses and low strains. Iterative techniques are also used for higher strains, where non-linear stiffness and non-viscous damping must be considered. The latter case does not appear to be commercially viable, except perhaps in the most specialised fields such as air-springs, which are beyond the scope of this paper. In general, viscous approximation appears to give satisfactory results and, using the computer, the designer can obtain reliable comprehensive data.

A quick appraisal of this brief summary reveals two principal regions where the designer cannot make adequate simple estimates. These are the stress and strain analyses of complex shapes, particularly those involving metal parts within the rubber and the analyses of heat generation and temperature distribution within a part, usually as a result of dynamic excitation. Both these factors are important in determining the life of a bonded unit, and it is on these that the bulk of this paper concentrates. First, however, the use of the computer in a more general application problem will be described.

11.2 THE DESIGN OF A MARINE ENGINE MOUNTING SYSTEM

This typical application for rubber/metal components presents no theoretical problems of any consequence at the relatively low deflections involved. All the classical, linear stiffness, viscous damp-ng parameters and small angle theories can be successfully implemented to provide the bulk of the designer's requirements. The calculations involved were hence those transferred to the digital computer quite early in many industrial computer developments to replace the laborious, incomplete, hand methods [8].

A system implementation from a few years ago has now produced a useful feedback of service data against which the calculated performance has been compared. The installation involved was that of the engine mounting system for the St Edmund, a British Rail car ferry running between Harwich and the Hook of Holland. We were involved with the builders, Cammel Laird and the Dutch design consultants, Technische Dienst TNO-TH, and attention was focused on providing passenger and crew comfort with respect to both noise and structural vibration. Because of previous success with other ferries, it was decided at an early stage to rigidly mount the gear box and flexibly mount the engine.

The flexible mounting system thus had to be the usual compromise between differing requirements. It had to be stiff enough to prevent large deflections under ship roll and pitch, so as to protect the flexible shaft coupling and the flexible connections to the exhaust, and soft enough to give good isolation of out-of-balance and torque effects. The constraints placed upon us were such that the critical resonant frequencies of the system had to be made to lie either below 3 Hz, between 4·6 and 6·7 Hz, or between 9·2 and 12·5 Hz. Noise transmission dictated a maximum of below 9·2 Hz and seaway movements were such that to use mountings giving below 3 Hz resonant frequencies would produce a dangerously flexible system. Thus the 4·6 to 6·7 Hz range had to be utilised. The mounting arrangement had to be chosen to produce the most easily excited natural frequencies in this band, and in order that we could propose a system it was necessary, given the required ship and engine information, for these frequencies to be calculated for a number of varying design systems. The main information required related to engine mass and inertias, angles of roll, pitch, etc., plus the service rotational speeds. Given these parameters the program used calculated static deflections and forced vibration amplitudes at any desired point on the suspended mass, with phase angles and transmitted forces, as well as the six natural frequencies related to coupled lateral, longitudinal vertical, roll, pitch and yaw modes of vibration. For this particular application, the least critical of these were the nearest to longitudinal and pure roll conditions.

158 A. J. Reed and J. Thorpe

Additional information required in these calculations related to the rubber properties, dynamic and static stiffnesses and damping coefficients, as well as actual mounting positions and mounting inclinations.

Taking into account many more details such as buffer clearances, creep and misalignment corrections, a theoretically acceptable mounting system was recommended using large 'sandwich' mountings without interleaves. This system was eventually tested with an actual engine, and predicted frequencies and movements found to be extremely close to those measured. Subsequent ship-board performance has been good.

11.3 DESIGN DECISIONS BASED ON STRESS LEVELS CALCULATED BY FINITE ELEMENT ANALYSIS

The use of finite element methods has increased considerably over recent years. The method is very dependent on the digital computer, except for the smallest, simplest models, as the calculation involved, in general, is extremely repetitive and laborious. The advent of electronic digital computers allowed development of finite element programs and at the same time generated considerable activity in the development of basic and eventually more complex and useful elements. A generally useful program must have a good library of available elements and data handling systems. It is thus, over the last few years, that a number of large and powerful programs have become commercially available, to be bought, rented or accessed through time-sharing systems.

The fundamental concepts of the theoretical method will not be discussed here. There is now a vast literature on the subject in general, mainly directed towards the metallic and generally 'stiff' materials used in engineering, Rockey et al. [7] giving a good basic exposition of the method. It is not difficult for those with the necessary knowledge and facilities to write their own program and, in the case of the more specialised aspects, it is probably necessary to do this, using one or more specially developed elements. This certainly applies to the accurate representation of high strains in rubber [6]. The number of man-hours involved is likely to be very high, so that for industry, in general, commercially available programs must be considered, particularly when the structures involved necessitate the use of a number of different element types.

A discussion should now be given to demonstrate how use can be made of a commercially available program to obtain comparative design assessments for the stress levels in a single bonded unit under static load. As rubber is considered a difficult, if not dangerous, material to analyse in such a way, some justification for the method is presented first, before the analysis of a difficult shape, where primary interest lay in the stress levels of metal interleaves embedded in the rubber. The justification takes the form of the compression of a rubber sandwich to show how some confidence in the rubber modelling was obtained. It must be emphasised, however, that, to date, this approach is primarily used to compare designs at low strains, not to produce absolute stress levels. One of the main reasons for concentrating on comparison is the difficulty in determining a sufficiently accurate value for Poisson's ratio in a given rubber application. The program used for the work described, is based on classical elastic theory with linear Hooke's law stress–strain relationships. The elements used in the main analysis were, however, relatively sophisticated, being twenty-node, three-dimensional solids with parabolic curve fitting for the deflections of the three nodes on each defining edge.

11.4 MODELLING RUBBER IN COMPRESSION

The part chosen for examination consisted of seven layers of rubber between eight metal plates. The cross-section of the part was circular, so that one rubber section was effectively a flat circular disc, with each flat face bonded to a metal plate. In compression, along the cylindrical axis, an axi-symmetrical problem was produced. This was considered appropriate for a trial analysis of rubber parameters as computer time and costs were hence reduced from the general three-dimensional levels.

The important assessments required were primarily related to the forces produced for a given deflection as the immediate future analyses were expected to be focused on metal interleaf stresses. Thus the accuracy with which the rubber transmitted forces to the interleaves was of paramount importance. The actual model used consisted of one disc of the seven for the whole circular part as shown along one radial section in Fig. 11.1. The nodes, for the axi-symmetric problem are actually rings and each element a solid of revolution produced by rotating the shown cross-sections around the y-axis. Each element has eight nodes, four corner and four mid-sided.

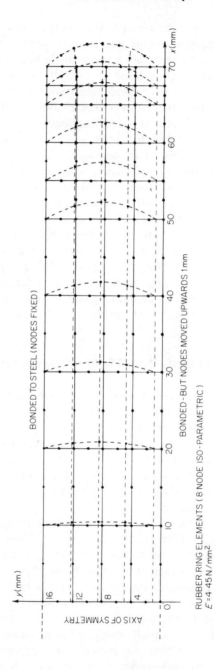

FIG. 11.1. Geometry of model before and after deflection.

They are iso-parametric, parabolic deflection elements using Gauss-point stress and strain calculation. The element stiffness matrix is calculated by Gaussian numerical integration of an order chosen according to application. In this case, the order is three, whereas for steel elements, at least, order two is satisfactory. The detail of the theory behind these specialised elements is contained in various program manuals and in one of the original books on the subject by Zienkiewicz [9].

Although only half the Fig. 11.1 model need be analysed, as the centre plane must move upwards 0·5 mm parallel to the outer planes (by symmetry), the whole section is shown. The remaining details are given on the diagram. The material properties shown relate to 60°IRHD hard rubber at low strain.

Figure 11.2 shows the variation of nodal point strain in the three coordinate directions along the radial section at the outer and central planes. Variation near the free, outer, bulging surface is rather too rapid for full confidence. The model used is relatively simple and more local details would require a finer mesh. Figure 11.3 shows a similar variation, but through the vertical thickness. Figure 11.4a gives the forces necessary to provide the given deflection from element to element. The points are plotted at the element centres in a radial sense. The total force is 42 800 N for the 1 mm deflection on the section or 7 mm deflection on the part. Nodal forces within each element oscillate around ±25 per cent from mid-side to corner nodes with the mesh shown. As the forces are totals for the complete element rings, the average y-direction force per unit area is found by dividing by the ring area. This gives the graph of derived normal stress also shown (Fig. 11.4) which averages very closely to the mean pressure on an interleaf plate. The smooth parabolic reduction from the centre has been experimentally verified, but the small local rise at the outer edge has to be viewed with an open mind.

Figure 11.4b gives a typical force–deflection curve for this part in the 60 H rubber. The tested force for 7 mm is a little greater than 40 000 N and the curve is sufficiently linear even beyond the deflection used. Figure 11.5 gives spacial representation of the nodal Von Mises and shear stresses. The former is used mainly as a fatigue criterion in metals, being a measure of principal stress differences in the three directions and given by f, where

$$2f^2 = (f_1 - f_2)^2 + (f_2 - f_3)^2 + (f_3 - f_1)^2 \qquad (11.1)$$

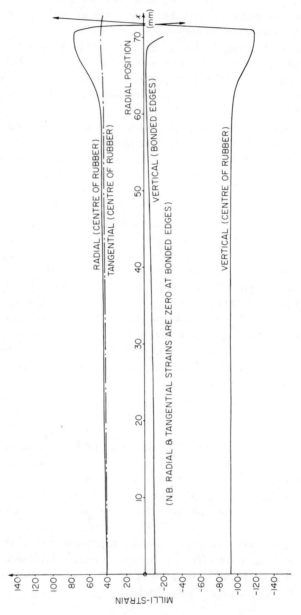

Fig. 11.2. Nodal strains. Radial variation.

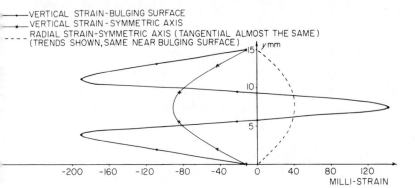

FIG. 11.3. Nodal strains. Through the rubber thickness.

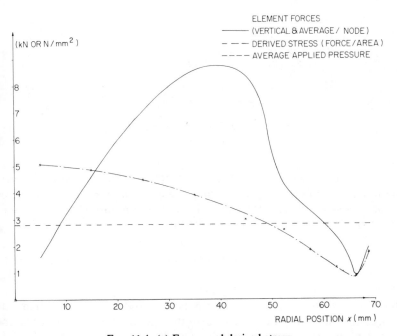

FIG. 11.4. (a) Forces and derived stress.

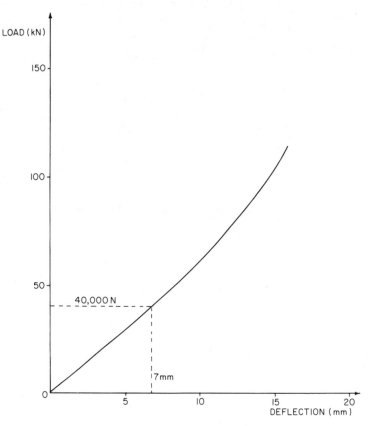

FIG. 11.4. (b) Tested force–deflection.

for principal stresses f_1, f_2 and f_3 at a point. The shear stresses ideally should be zero at the surfaces where Gauss point averaging and other errors show non-zero values. Similarly, the contours should run into the corners where a singularity probably exists.

Figures 11.6 and 11.7 show the calculated stresses given by the program which are double dependent on the classical formulae using Poisson's ratio, here taken to be 0·499. Although generally smooth, the nodal figures oscillated to some extent in the y-direction owing to the linear y-direction strain interpolation, and so average element figures are shown. These less accurate figures though appearing to give the correct variations show an absolute level around 50 per cent

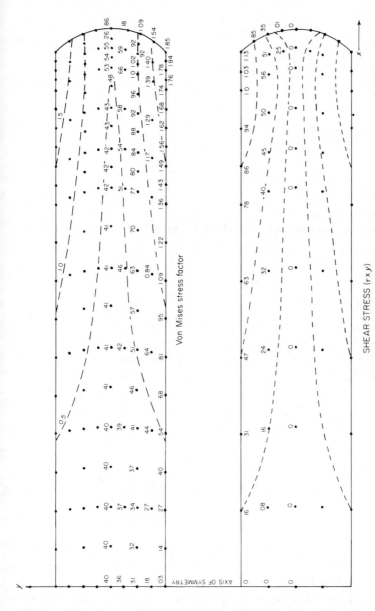

FIG. 11.5. Nodal stresses showing some equi-stress lines (N/mm²).

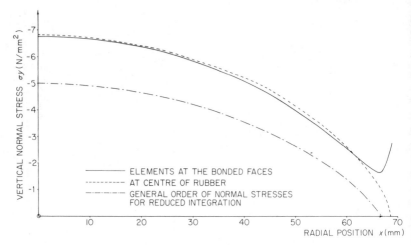

FIG. 11.6. Average element normal stress. Vertical (direction of compression).

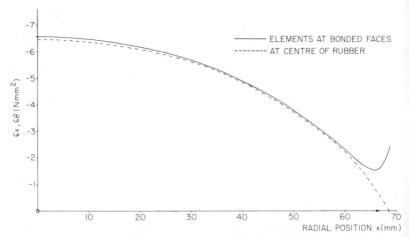

FIG. 11.7. Average element normal stress. Radial and tangential (a small difference).

higher than expected (*cf.* Fig. 11.4a). The value of Poisson's ratio is probably the main factor here, but as other analyses have also shown, a value of 0·499 gives good force–deflection results.

A more recent analysis using reduced integration gave forces and stresses very close to the known accurate figures and reduced the oscillation, all nodal variations being as shown by the chain dotted line in Fig. 11.6. Overall, the results were very promising and so encouraged progress to be made in more complex applications. If costs could be balanced effectively against the need, more elements in the y-direction would obviously improve accuracy.

11.5   APPLICATION OF THE PROGRAM TO GENERAL THREE-DIMENSIONAL ANALYSES

The results shown in the previous section suggested that forces would be accurately transmitted through rubber-to-metal interleaves if the iso-parametric elements were employed in at least two layers along compression directions. The average small strain assumption also had to hold, unless iterative loading and geometric deformation was to be attempted. The non-linear nature of rubber behaviour as strains increased was also considered to be an acceptable factor for the lower strains used, particularly as detailed strain concentration in the rubber was not the primary target.

The main objective has been, and at present continues to be, the comparative design assessment of various configurations with respect to metal part stress. Further development of the commercial programs, both theoretically and in their use, is taking place and more confident, absolute assessments may soon be possible. High strain theory has also been allied to finite element methods in the rubber industry, and applications have been found in this field of rubber–metal bonded units. Prediction of the stiffness of the circular rubber sandwich already discussed has also been successfully accomplished using a high strain finite element program.

For general three-dimensional analysis, the iso-parametric element approach using the same program which gave the results depicted earlier, appears to be the best available tool, at least within the limits described above. A particular analysis of a multi-layer 'CHEVRON' shaped mounting, where shear and compression loading is applied, demonstrates a typical finite element program usage.

The designers of axle-box, primary suspension units on rail vehicles require the three main stiffnesses along transverse, vertical and longitudinal directions to have ratios which are difficult to obtain in a limited width with conventional units, such as plane sandwich mountings. A solution has been produced by, in effect, bending the above type of mountings into a 'V' shape. This produces relatively higher transverse stiffnesses, for example. This type of part, called a 'CHEVRON' and illustrated in front and side elevation by Figs 11.8 and 11.9a, is used in a 'V' arrangement with another matched part, such that under vertical loading, each unit experiences shear and compression. Often the resulting bending of the interleaves is sufficiently critical that the metal stresses are more important than the rubber stresses. A good design produces an effective balance. Many such parts have, in fact, successfully lived through many years of arduous service conditions. A typical configuration in service position is shown by Fig. 11.9b. The diagrams show the main problem in analysing such a part. This problem is the lack of symmetry and the necessary analysis of all layers. The whole part is simplified in one way only, by dividing it into two, through the apex plane, the section of which is shown in Fig. 11.8. The loading is such that no deflections normal to this plane can occur for points on the plane. Another complication not shown by the diagrams is that the lower two rubber sections have large corner pieces removed, leaving a more oval section when viewed from the Z-axis towards the origin. This feature is thought to improve stresses in the lower plates. Overall, a very large number of nodes and relatively, elements, have to be used even with the 20-node iso-parametric type which help to model the curved faces of the part. The theoretical background of the analysis is as indicated earlier.

As shown in Fig. 11.8 the upper rubber face, in reality bonded to outer metals, is fixed and the lower plate deflected such that compression and shear occur, with an overall angle of movement in the ZX plane of 11°. The actual deflection used is only relative, but as before, is such that less than 10 per cent average strain occurs. To maintain separate rubber and steel material properties right up to the boundary faces, each layer of rubber and steel is modelled separately, being joined by doubled coupled nodes in identical positions. This produces meaningful boundary results which would otherwise be averaged to lose significance. The rubber properties used are as in the previous example with Young's modulus for steel being taken as $206\,280\,N/mm^2$, shear modulus $96\,200\,N/mm^2$ and Poisson's ratio $0·28$. The steel elements incorporate reduced (2nd order) integration.

Three analyses of 'CHEVRONS' are depicted in the Von Mises stress plots given by Figs. 11.10–11.18. The figures given are the values at each node scaled to produce directly comparable relative

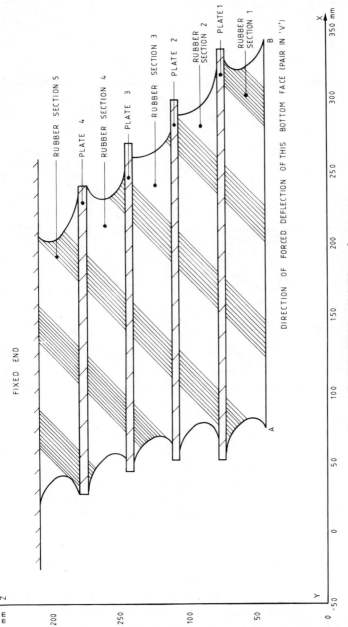

FIG. 11.8. CHEVRON model. Plane of symmetry.

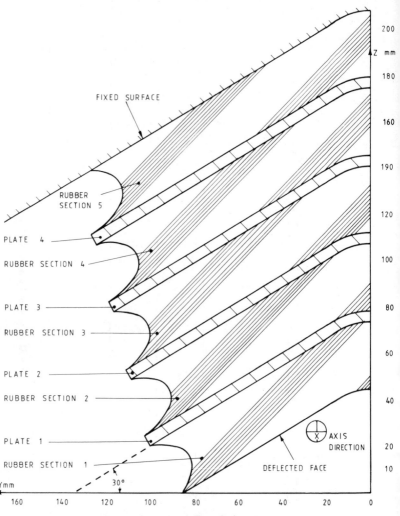

FIG. 11.9. (a) Undeflected view YZ-plane.

values for all three sets of results. The absolute figures have little meaning. Figures 11.10 to 11.13 show the factor for the undersides of plates 1 to 4, respectively, of the basic standard design. The plates are largely in a state of tension on this surface with respect to stress, but strain results show a reduction of tension from a maximum at the

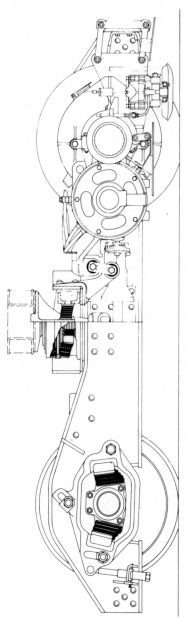

FIG. 11.9. (b) In-service position of CHEVRON.

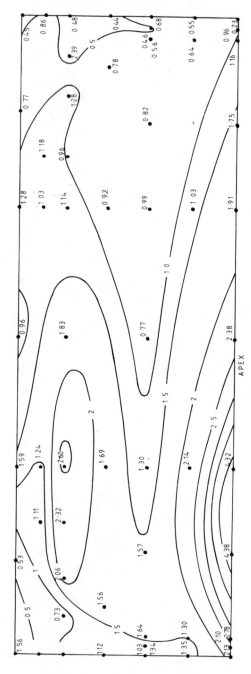

FIG. 11.10. Plate 1 (4·75 mm) underside stress factor.

ex, to small regions of compression near the outer, long edge. This
a very loose description of a complex, three-dimensional change
d so many results would have to be shown to give a clear picture,
at attention is given only to the fatigue relevant Von Mises factor.
he plates in these four plots are as shown 4·75-mm thick.

Figures 11.14 to 11.17 give exactly parallel results including equal
iffness, for the same part except that the plates are 6-mm thick. The
verage stresses are around 5 per cent less than for the original
esigns with a similar order reduction for maximum stresses at the
pexes of plates 2 and 3 (Figs. 11.15 and 11.16, respectively).

The remaining Von Mises plot (Fig. 11.18) is for the underside of a
ouble thickness central plate in a very similar design, but having
nly four rubber sections. The other two interleaves are on either side
f the thick one and are 6-mm thick. Because the thicker plate was
sed, it was hoped to reduce stresses as the first comparison suggests,
ut using the same overall height of 'CHEVRON' (Z-direction, Fig.
1.9) the four individual rubber sections were thicker, thus increasing
he geometric effect which causes the plate bending (off-set pressure
entres). The results indicate a cancellation of effects with very
imilar stress levels and greater concentration of maximum stresses at
he thick plate apex.

Further development on the 'CHEVRON' design is continuing with
number of smaller, localised analyses, where various design factors
re being studied with the aim of reducing the interleaf stresses.

11.6   CALCULATION OF TEMPERATURE RISES OWING TO DYNAMIC HEAT GENERATION IN RUBBER

ertain rubber-bonded-to-metal parts have thick rubber sections with
n application which produces considerable dynamic heat generation,
wing to the internal damping where mechanical energy is lost in the
orm of heat energy. For thin rubber sections, bonded to metal, as in
he parts described previously, the heat is conducted away quite
uickly and no problem is produced. However, thicker rubber sec-
ions in such applications as vibrating screens experience large tem-
erature rises.

Classical analyses have been produced for the most representative
hape, the cylinder, but the problems of dealing with composite
aterials and difficult boundary conditions have made this approach

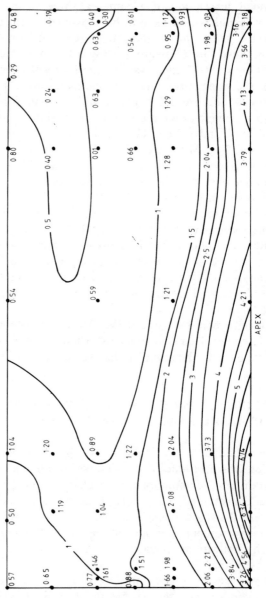

FIG. 11.11. Plate 2 (4·75 mm) underside stress factor.

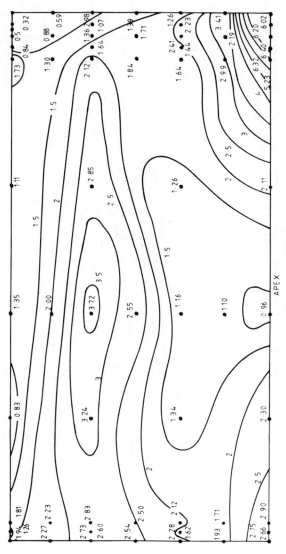

FIG. 11.12. Plate 3 (4·75 mm) underside stress factor.

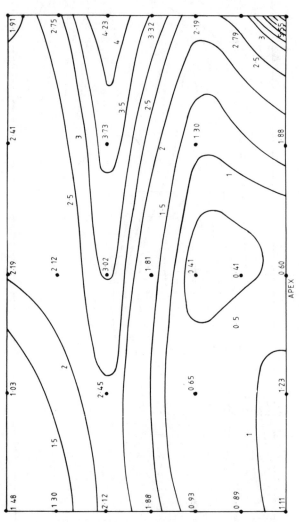

FIG. 11.13. Plate 4 (4·75 mm) underside stress factor.

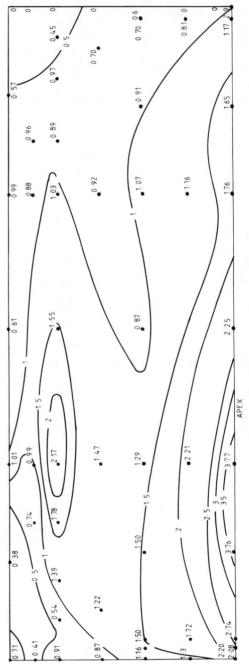

FIG. 11.14. Plate 1 (6 mm) underside stress factor.

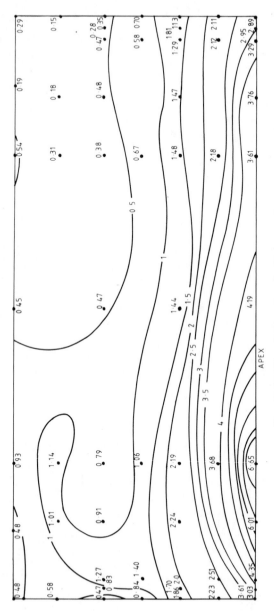

FIG. 11.15. Plate 2 (6 mm) underside stress factor.

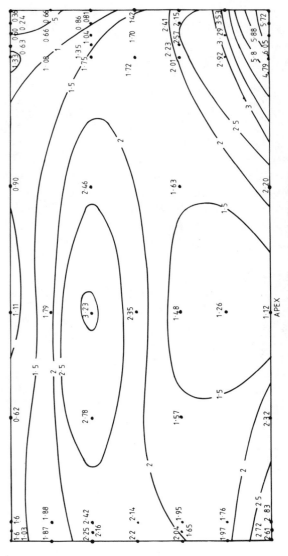

Fig. 11.16. Plate 3 (6 mm) underside stress factor.

A. J. Reed and J. Thorpe

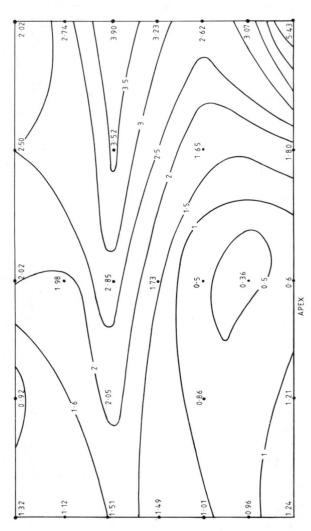

FIG. 11.17. Plate 4 (6 mm) underside stress factor.

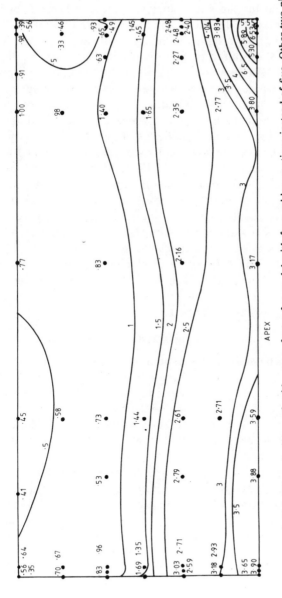

FIG. 11.18. Central thick (12-mm) plate underside stress factor for model with four rubber sections instead of five. Other two plates 6 mm.

too inflexible and laborious to utilise successfully. The finite element program used for the stress analyses, previously described, also ha elements designed for heat conduction and generation. These ele ments utilise the classical theory [10] but handle the varyin requirements of differing materials and boundary conditions ver successfully. The following exposition shows the basic approach for single part, where measured temperatures are available.

The part chosen was cylindrical, 150-mm diameter and 125-m high, along the cylindrical axis. Hence one-quarter of the part co stituted the finite element model as shown by Figs. 11.19 and 11.2(The metal plate bonded on the flat surfaces was assumed to be total insulated in the model (Fig. 11.19), and in test conditions a thic asbestos slab nearly generated this condition. At the curved bou daries, free conduction and radiation into still air at 29°C wa assumed. Test conditions were set up so that no significant draught affected the part without restricting the surrounding air. In the test th part was vibrated sinusoidally at an amplitude of ±2 mm wit frequency 13·5 Hz. The force producing this was monitored on th machine which also gave damping factors, both being digital reading from an analogue device in the machine. The resulting average he generation is given by the formula

$$Q = \pi F x \sin \phi \cdot W/V \qquad (11.2$$

where Q is the heat energy generation per second per unit volume; the force amplitude; x the deflection amplitude; ϕ the phas difference between force and deflection; W the frequency in cycle per second; and V the volume of the rubber.

This indicated a heat generation around 5×10^{-6} W/mm^3, a figur with a probable error of around 10 per cent from experience with th machine accuracy.

The other crucial quantities are the rubber thermal conductivity and total emissivity H (perhaps more commonly called the coefficie of surface heat transfer) into the given air condition. The first of the has been measured many times for the compounds used, and wa taken to be $2 \cdot 34 \times 10^{-4}$ W/mm·°C, a somewhat higher figure than mo text books appear to quote. The rubber used was a natural-base compound of a hardness around 50°IRHD. The emissivity H defined by the equation

$$K \partial V/\partial n + H(V - V_0) = 0 \qquad (11.$$

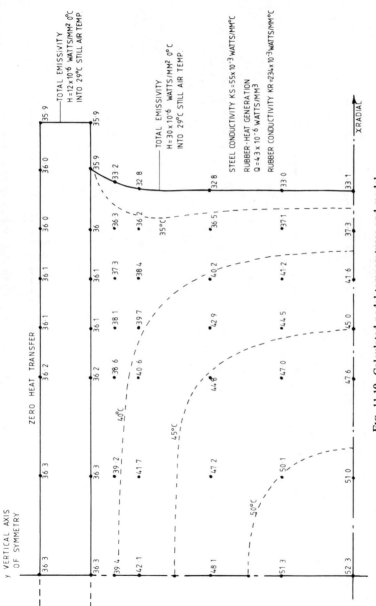

FIG. 11.19. Calculated nodal temperatures and model.

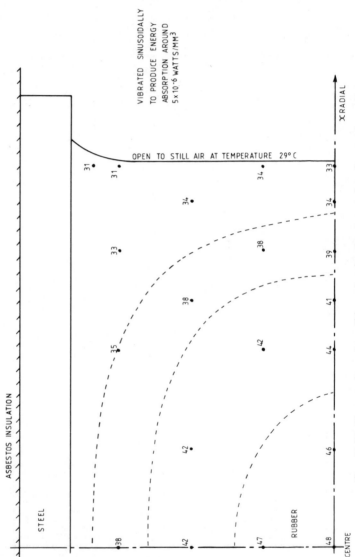

FIG. 11.20. Test results and conditions showing thermocouple point steady readings.

where $-K \partial V/\partial n \equiv$ heat flux crossing the surface; V is the inner, body temperature variable at the boundary point in question with n defining the normal to the surface at that point and V_0 the outer, medium temperature there. This quantity has been measured for many different conditions by electrical heating experiments as well as estimation from vibration tests, and a figure of around 20×10^{-6} W/mm²·°C was expected. Having completed preliminary tests the part was vibrated as described for around 8 hr, until no time variation of temperature took place, and previously positioned thermocouples were used to record the temperature distribution. These figures were used approximately in conjunction with the simple, sphere steady-temperature formula to assess H and check back to Q. The formula used is

$$V - V_0 = Q[(a^2 - r^2)/6K + a/3H] \qquad (11.4)$$

where V is the sphere temperature variable for heat generation per unit volume Q, conductivity K, general point radius r and outer radius a. From this, taking $V = V_1$, the central temperature $(r = 0)$ and $V = V_2$ the surface, body temperature $(r = a)$ the formula for H is

$$H = 2K/a \left[\frac{V_1 - V_0}{V_2 - V_0} - 1 \right] \qquad (11.5)$$

Thus using estimates for a, V_1 and V_2 from tests (Fig. 11.20), a good approximation for H can be obtained and a value of 30×10^{-6} W/mm²·°C was produced. Calculating Q from eqn (11.4) gave a value of $4 \cdot 3 \times 10^{-6}$ W/mm³. Although not accurate, these were close enough to the expected values for them to be used directly in the finite element analysis, giving the results shown by Fig. 11.19. The steel conductivity and total emissivity were taken as 55×10^{-3} W/mm·°C and 12×10^{-6} W/mm²·°C, respectively, using quoted figures and test results for the particular steel and surface condition in the part tested.

Comparisons of Figs. 11.19 and 11.20 show that differences are small, even though accuracy in some of the assumed coefficients was not high. In general, the results encourage the use of this method for heat analysis. It is worthy of note that another important area where applications for the method are found is in estimation of state of cure. The temperature distribution can be converted to 'cure units' using Arrhenius-type equations. Work is currently continuing in this field.

Another project under investigation is that of cooling-shrinkage stresses in bonded components.

11.7 CONCLUSION

Specific problems have been given detailed study where commercial pressures have demanded answers. The use of the digital computer has allowed previously difficult or impractical calculations to be carried out. The simulation of small strain dynamic problems, particularly in forecasting natural frequencies, has progressed entirely successfully.

In the more advanced finite element work, reasonable small strain, linear results have been obtained and some confidence is being built up to provide design comparisons. In general, apart from the examples described, many other external loading stress applications and a number of heat-transfer applications have been analysed.

In its application to rubber, the finite element method is still very much a research and development tool, and requires careful control and interpretation. The recent and the impending liability legislation and the normal competitive pressures are demanding better and more precise prediction of product performance, and finite element techniques may soon be in use in most design offices.

REFERENCES

1. P. B. LINDLEY. *Engineering Design with Natural Rubber*; *NR Technical Bulletin*. MRPRA.
2. A. R. PAYNE. '*Dynamic properties of vulcanised rubbers*; *5 shape factors and functions in rubber engineering*'. *RAPRA Research Report*, 1957, No. 84.
3. A. N. GENT and P. B. LINDLEY. 'The compression of bonded rubber blocks'. *Proc. Inst. Mech. Eng.*, 1959, **173**, 3, 111–122.
4. A. N. GENT and E. A. HEINECKE. 'Compression, bending and shear of bonded rubber blocks'. *Poly. Eng. & Sci.*, 1970, **10**, 48.
5. B. P. HOLOWNIA. 'Effects of Poisson's ratio on bonded rubber blocks'. *J. Strain Anal.*, 1972, **7**, 236.
6. P. B. LINDLEY. 'A finite-element program for the plane-strain analysis of rubber'. *J. Strain Anal.*, 1975, **10**, 25.
7. K. C. ROCKEY, H. R. EVANS, D. W. GRIFFITH and D. A. NETHERCOT. (1975). *The Finite Element Method.* Crosby Lockwood Staples.

. C. M. HARRIS and C. E. CREDE. (1961). *Shock and Vibration Handbook.* Vol. 1 (pp. 3–1 to 3–62); Vol. 2 (pp. 30–1 to 30–33). McGraw-Hill.
. O. C. ZIENKIEWICZ. (1971). *The Finite Element Method in Engineering Service,* 2nd edn. McGraw-Hill.
. H. S. CARSLAW and J. C. JAEGER. (1959). *The Conduction of Heat in Solids,* 2nd edn. Clarendon Press.

Chapter 12

ENGINEERING APPLICATIONS OF RESILIENT MATERIALS

D. E. NEWLAND and A. D. LIQUORISH

12.1 INTRODUCTION

The applications of resilient materials with which the authors have been concerned lie in the civil and mechanical engineering fields. They can be classified roughly into four groups according to the principal function of the resilient element in each case. These groups are where: (i) static load distribution is the main function; (ii) the absorption or control of vibration is the main objective; (iii) impact absorption is important; and (iv) a flexible closure of ducting or pipework is required.

In all these applications, the resilient element must permit relative movement of its adjacent parts. This may occur either occasionally or regularly, and the rate of load application may be gradual or extremely rapid. Flexibility is the key parameter and it must be achieved and maintained for an adequate lifetime under a possibly hostile environment.

Since the flexibility of elastomeric resilient elements of the type described in this paper depends very much on the rate of load application and the average static pressure, it is most important that adequate dynamic testing has been completed before a new material is used in applications which involve dynamic loads and/or high static pressures [1, 2, 3]. For most applications, it has been found sufficient for the mechanical properties of a material in the form of a solid bearing to be described by: (a) the static load versus deflection characteristic under normal (generally compression) loading and in shear; (b) the long-term creep behaviour in the form of a graph of pad deflection against time under constant static (and, if necessary,

189

dynamic) loading; and (c) the dynamic modulus and loss factor at different frequencies and amplitudes of vibration. These dynamic properties are generally derived from measurements of the natural frequency and damping of a test system involving a rigid block mounted on one or more bearings whose lower surface is against a (relatively) immovable foundation.

With this basic minimum information, satisfactory predictions of the behaviour in service of a wide range of different resilient elements have proved possible, and some of these applications are now described.

12.2   PRINCIPAL FUNCTION OF LOAD DISTRIBUTION

The absorption of movement owing to expansion and contraction occasioned by changing temperature is a fundamental design requirement in many large structures, and rubber mountings for bridges have been used for many years (Fig. 12.1). Lateral movement is accommodated by shear of the resilient element, which must have the flexibility necessary to permit this movement while carrying

FIG. 12.1. Typical rubber/steel bridge bearing *in situ*.

ignificant compressive stress due to the supported weight. Modern constructional techniques are now making increasing use of resilient seatings in large buildings, where they have two main functions: (i) to absorb thermal movements and (ii) to eliminate high stresses at the junctions between sub-structures and main supporting beams. Additionally, resilient seatings serve to prevent corrosion, fretting and spalling, to prevent the ingress of moisture, and to act as a barrier against noise transmission (Fig. 12.2). The type of material and thickness necessary will depend on the horizontal and rotational

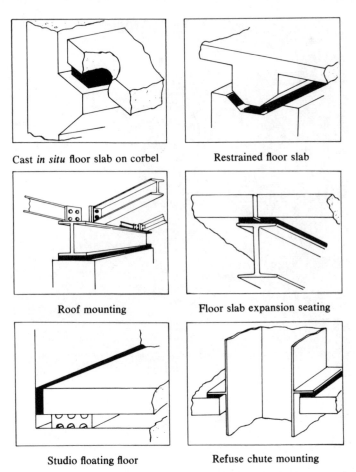

Cast *in situ* floor slab on corbel   Restrained floor slab

Roof mounting   Floor slab expansion seating

Studio floating floor   Refuse chute mounting

FIG. 12.2. Examples of the application of resilient seatings.

movement that the seating must absorb, together with the irregularity
and lack of parallelism of the two adjacent surfaces. Where necessary
very large structural movements can be accommodated by in
corporating a sliding surface with the resilient seating (Fig. 12.3).

A recent building that has made extensive use of resilient seating
is the Dubai International Trade and Exhibition Centre [4]. Thi
includes a 34-storey central tower in reinforced concrete, the oute
main structural columns of which are designed to form part of the sur
screen for the perimeter windows (Fig. 12.4). With extremes o
summer temperature from 8 to 48°C, and the sun travelling practically

FIG. 12.3. Components of a sliding resilient seating to accommodate expansion in a
viaduct on the Shoreham bypass (the A27).

FIG. 12.4. The Dubai International Trade and Exhibition Centre under construction. (Courtesy of Bernard Sunley & Sons.)

over the top of the building every day, differential vertical movements of the order of ±40 mm were predicted between the inner and outer vertical columns of the building. Such large movements could not be tolerated for a standard concrete-to-concrete connection between the floor structure and the vertical columns because of the relative rotation occurring as the floor tilts slightly first to one side of the building and then towards the other side as the sun passes overhead.

To accommodate this movement, two types of resilient seatings have been incorporated in the structure. High load bearing seatings capable of withstanding a continuous static stress in compression of 8.5 MN/m^2 are mounted under the main floor beams at their junction with the vertical supporting structure, and seatings of a softer material, capable of withstanding about 2.0 MN/m^2 are positioned to provide continuous support round the periphery of the floor slabs (Fig. 12.5). Both types of seatings are of polychloroprene modified by the inclusion of cellular particles, and the high load grade is of laminated reinforced construction, having alternate plies of high-strength synthetic fabric moulded into the bearing material.

The construction of the Dubai building involved casting the floor slabs *in situ*, and the procedure followed was to secure the resilient seatings in place with contact adhesive and then to tape non-load bearing void filler strips on either side of the seating to prevent the ingress of wet concrete. This procedure has now been followed for a range of applications, with complete success, and no specialised constructional techniques have been required by the building contractors.

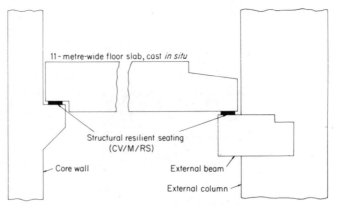

FIG. 12.5. Dubai tower block floor slab, simply supported on resilient seatings.

The function of resilient seatings in reducing high contact stresses is important in railway applications where resilient pads are placed between the rails and sleepers. Concrete sleepers, particularly, require a resilient interface layer if localised crushing and spalling of the concrete surface is to be prevented (Fig. 12.6). A variety of different rail pads has been in extensive use for many years. Their design involves balancing the need for sufficient flexibility to give an even load distribution against the requirement that rail deflection under load must be limited to keep rail stresses to acceptable levels and prevent slackening of the fastening clips (Fig. 12.7).

The same requirements apply for the design of resilient crane rail seatings where axle loads are generally much heavier [5]. As a result, the crane rail is supported continuously along its length on a steel or concrete foundation, and the resilient crane rail strip runs continuously along the undersurface of the rail (Fig. 12.8). The resilient layer compensates for the possibly uneven surface of the support structure, prevents fretting corrosion, and reduces the transmission of noise and vibration from the crane rail, as well as acting to distribute the high wheel loads evenly and minimise high local stresses in the supporting foundation. This is especially important when the crane rails are

FIG. 12.6. Concrete railway sleepers with resilient rail pads and fastenings before assembly.

FIG. 12.7. Assembled track incorporating rail pads.

carried above ground on welded gantry girders, when high local stresses can cause fatigue and fracture of the web-to-flange welds.

A detailed series of tests to measure the strains in a loaded gantry girder under a crane rail was carried out by the authors some years ago, and a typical result is shown in Fig. 12.9. Strain gauge measurements were taken at two locations, chosen because the crane girders were identical at each location, but resilient crane rail strip was fitted at only one of these locations. The same 30-tonne crane operated over both locations. The graph in Fig. 12.9 shows the strain recorded in each case in a strain gauge placed just below the top flange of the girder as the crane was slowly moved across the measuring point. The gauges were aligned as shown in the inset view and compressive strain is recorded. The graphs show two peaks in each case as the two wheels of the crane bogie passed over the measuring point. It can be seen that the localised strain in the web close to the top flange of these 195-mm deep crane girders was about 50 per cent greater at the location without crane rail strip.

The compressive stress transmitted by a crane rail resilient strip depends on the flexibility of the seating and on the stiffness of the rail. Figure 12.10 shows crane rail strip (c.r.s.) stress (plotted

FIG. 12.8. A typical crane rail installation with continuous resilient crane rail strip in position under the rail.

horizontally) against crane wheel load (plotted vertically) for a typical general-duty resilient strip of 5·5 mm thickness under different weight rails. For a heavy rail section (164 kg/m) the c.r.s. pad stress is much less than for a light (33 kg/m) weight rail. Figure 12.10 also shows, as the intersecting curves, the calculated maximum bending stress in the rails. Each of these curves crosses the graphs of crane wheel load against c.r.s. pad stress at the value of crane wheel load which causes the stated value of rail bending stress.

The choice of an acceptable maximum permissible level for rail bending stress depends on the strength of the welded joints between rails, rather than on the strength of the rails themselves, and the probable fatigue strength level under repeated loading as the crane

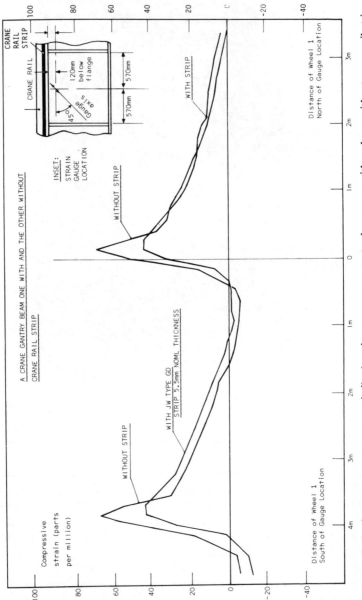

Fig. 12.9. Strain measurements at two similar locations on a crane gantry beam, one with and one without crane rail strip.

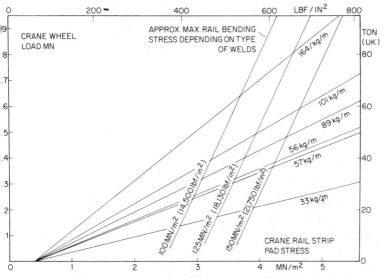

FIG. 12.10. Relationship between wheel load and crane rail strip stress for different crane rails (for a typical resilient crane rail strip).

moves backwards and forwards may be relatively low. The limiting design criterion will therefore usually be rail stress rather than the crushing stress of the c.r.s. pads. The resilient seating material used is generally of a laminated construction with layers of cork-modified synthetic rubber compound reinforced by synthetic rubber proofed fabric. In addition to providing the correct load–deflection characteristic, it is important that the material also has a high recovery property after load removal to reduce the possible tendency towards slackening of the rail fastenings under continuing use.

Another mechanical engineering application is the provision of resilient pipe supports and guides to accommodate movement arising from thermal expansion, thereby preventing problems of pipe distortion and over-stressing as well as problems of fretting corrosion caused by small local movements of the support structure (Figs. 12.11 and 12.12).

12.3 PRINCIPAL FUNCTION OF VIBRATION CONTROL

The concept of isolating a system from vibration by mounting it on springs or of using a sprung suspension to limit the vibration emanating from machinery has been understood for many years. Moulded

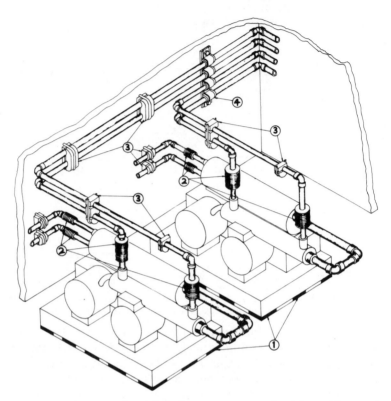

FIG. 12.11. Arrangement of a typical pipe support system involving resilient seating
1, Resilient anti-vibration pads; 2, expansion bellows; 3, resilient clamp blocks;
resilient clip strip.

rubber spring mountings are widely used, and cheaper machinery
mounting pads of cork-modified rubber compound have found ex
tensive application (Figs. 12.13 and 12.14). The design of most c
these mountings is usually intended to achieve natural frequencies c
the mounted assembly on its suspension which are appreciably belo
the principal frequencies of excitation.

When the mounted assembly is itself rigid in comparison with th
flexibility of the mountings, approximate calculations of the mounte
results can be achieved without difficulty. There is more difficult
when the mounted assembly is itself somewhat flexible, and th
provision of a flexible suspension can, if not properly designed
accentuate rather than attenuate a vibration problem. These desig
problems became important as the trend developed for mounting large

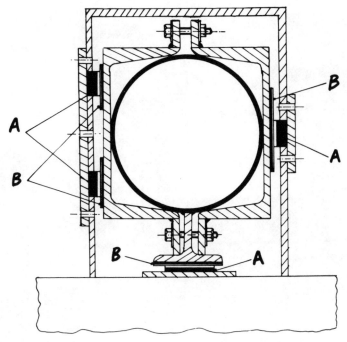

FIG. 12.12. Pipe support design involving nylon/PTFE sliding interfaces with the PTFE surface bonded to a resilient seating. A, PTFE-faced resilient pad; B, high-impact nylon surface.

and larger assemblies, and eventually whole buildings, on springs.

Another example of the control of vibration within a structure is the floating floor system employing two floors separated by an air space, with the upper floor supported on resilient mountings to reduce the transmission of vibration and acoustically decouple one mass from another (Fig. 12.15). This practice is in common use for the construction of television and recording studios, music practice rooms and anechoic chambers as well as for the construction of complete floors of upper storey plant rooms. Since the success of such a system depends on having a sufficiently heavy floating floor, it is normal practice to employ a reinforced concrete slab suspended on the resilient mountings. The thickness of the slab and the stiffness of the mountings can be adjusted to achieve the required degree of isolation.

Where the ambient noise level must be kept particularly low, such

FIG. 12.13. Refrigeration equipment supported on moulded rubber anti-vibration mountings. (Courtesy of Hall Thermotank International Ltd.)

as in a broadcasting studio, it is normal practice to design the system as a 'box-within-a-box', supporting the inner walls and the false ceiling on the isolated floor slab (Fig. 12.16). It is then also important to ensure that the cavity between the isolated room and its supporting structure is partially filled with a suitable sound absorbing material in order to prevent the possibility of acoustic resonance within the cavity.

The contribution to this volume by Professor Grootenhuis (Chapter 13) deals specifically with problems of resiliently mounted structures and therefore only one representative example of recent practice in this area will be given here. The 95 luxury flats being constructed in Ebury Street, London, provide a particularly good example of a complex building which has been separated from its foundation by a resilient layer designed to isolate the building from the ground transmitted noise and vibration emanating from tube trains running

FIG. 12.14. Air compressor supported on adjustable height mountings incorporating cork-modified rubber pads. (Courtesy of British Home Stores Ltd.)

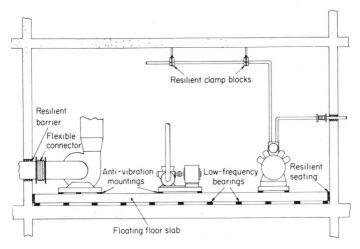

FIG. 12.15. Arrangement of a typical air-conditioning plant installation on the upper floor of a high building.

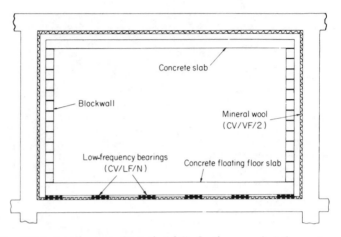

FIG. 12.16. Arrangement of a typical fully-floating room isolation system.

underground nearby (Fig. 12.17). The 24 000-tonne structure is supported on 1400 high-load non-metallic reinforced polychloroprene bearings 103 mm thick, which are designed to provide a vertical natural frequency of the order of 10 Hz at a stress of 5·6 MN/m². Although the resilient bearings are designed to withstand stresses several times greater than the operating conditions, fail-safe devices have been incorporated in accordance with the recommendations of BSI DD47 *Vibration Isolation of Structures by Elastomeric Mountings*.

Problems of isolating buildings from railway noise are the subject of continuing research, and although good results have been achieved in predicting the transmission of low frequency vibration, it has not yet been found possible to satisfactorily predict high frequency noise levels, probably because localised resonances cause significant local variations within a building [6]. Where new railway lines have been constructed in heavily populated areas, floating slab isolation systems have proved helpful in reducing ground excitation (Fig. 12.18). The basic design approach is to mount the track on a rigid beam structure which is itself 'floated' on resilient bearings, and the results achieved at the Barbican site in London and on the Piccadilly-line extension of London Transport to Heathrow airport have proved very satisfactory.

The same considerations apply to the design of a resilient foundation for forging hammers (Fig. 12.19), where intermittent bursts of severe ground vibration may occur as each blow falls. Hammer

FIG. 12.17. Resilient structural bearings shown in position during construction of luxury flats at Ebury Street (London) in 1977. (Courtesy of Ove Arup & Partners.)

foundation design has developed on a largely empirical basis over many years, but the recent increased emphasis on reducing environmental noise and vibration levels has encouraged a greater attention to reducing ground disturbance during hammer operation. A typical large modern hammer may have a blow energy at impact of some 50 tonne metres, the workpiece being supported on a cast steel anvil of some 450-tonne weight, and the anvil itself being carried on a reinforced concrete foundation block of perhaps 1500-tonne weight. The initial instantaneous downwards velocity of the foundation block as a result of hammer impact is typically about 0·1 m/s and the gross peak dynamic force transmitted by the foundation block to the surrounding soil may be some 4 times the static weight of the hammer and foundation if the foundation block is not mounted on a sprung

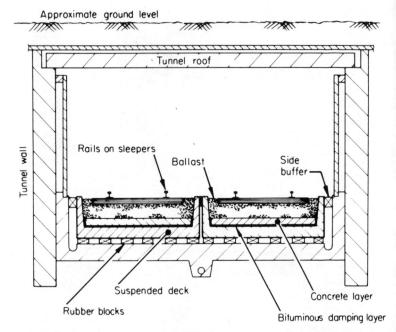

FIG. 12.18. Typical arrangement of resilient supports for railway floating slab track construction.

suspension system. To reduce the severity of this foundation block-to-soil interaction, resilient mountings of bonded cork material are often employed on the floor of the foundation pit below the concrete inertia block.

Improved isolation can be achieved by using thicker layers of softer material, and the pressures involved are low (about 0.3 MN/m^2), so that virtually any degree of isolation can be achieved by increasing the flexibility of the resilient layer sufficiently. Although elaborate fully sprung suspension systems have been used in a few critical applications, the normal current practice is to employ resilient mountings approximately 50 mm thick of total area equal to one-half the inertia block base area. Depending on ground conditions, this gives a vertical natural frequency of the foundation assembly of about 17 Hz and reduces the ground disturbance by a factor of about 2 compared with a system with no resilient layer.

Damping is particularly important in a recent new and unusual application of a resilient seating employed to prevent the self-excited

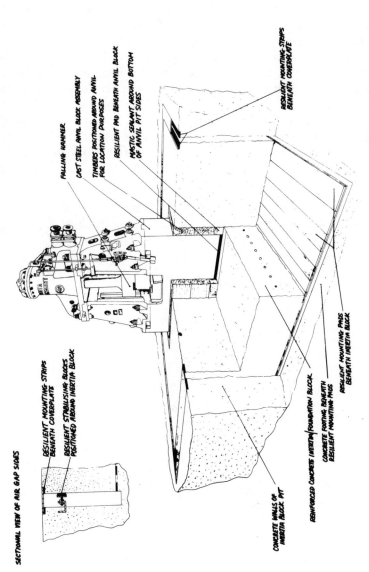

SECTIONAL VIEW OF AIR GAP SIDES

RESILIENT MOUNTING STRIPS
BENEATH COVERPLATE

RESILIENT STABILISING BLOCKS
POSITIONED AROUND INERTIA BLOCK

CONCRETE WALLS OF
INERTIA BLOCK PIT

REINFORCED CONCRETE INERTIA/FOUNDATION BLOCK

CONCRETE FOOTING BENEATH
RESILIENT MOUNTING PADS

RESILIENT MOUNTING PADS
BENEATH INERTIA BLOCK

FALLING HAMMER

CAST STEEL ANVIL BLOCK ASSEMBLY

TIMBERS POSITIONED AROUND ANVIL
FOR LOCATION PURPOSES

RESILIENT PAD BENEATH ANVIL BLOCK

MASTIC SEALANT AROUND BOTTOM
OF ANVIL PIT SIDES

RESILIENT MOUNTING STRIPS
BENEATH COVERPLATE

FIG. 12.19. Layout of a typical hammer installation with a resiliently mounted foundation block.

oscillation of factory chimneys due to wind forces. Tall steel chim
neys are prone to lateral swaying oscillations, brought about by th
regular shedding of vortices alternately from one side of the chimne
and then the other side under steady wind conditions. The problem
worst for welded-steel chimneys, which have very low inhere
structural damping, and one solution is to fit helical spoilers to the to
section of the chimney to break up the regular shedding of vortice
An alternative and cheaper approach, which has apparently bee
completely successful in a number of recent applications, is to i
corporate a resilient layer at the junction between the steel mountin
stool of the chimney and the concrete foundation block to which th
stool is bolted (Fig. 12.20). Provided that the resilient layer introduce

FIG. 12.20. A 30-m high welded-steel factory chimney mounted on a resilient seating
increase structural damping.

ufficient damping, the overall damping present in the chimney-
oundation assembly (Fig. 12.21) is increased to a level at which
elf-excited wind oscillations do not build up, the rate of energy
issipation by damping exceeding the rate of vibratory energy supply
rom the wind.

The design of resilient seatings for chimneys is a continuing subject
f research and development, with some of the tests taking place at
he University of Loughborough, and a delicate balance has to be
truck between providing sufficient flexibility in the resilient layer for
he 'working' of the resilient material to be sufficient for it to dissipate
nough energy, while at the same time providing sufficient stiffness to
top the top of the chimney swaying too far under gale conditions.

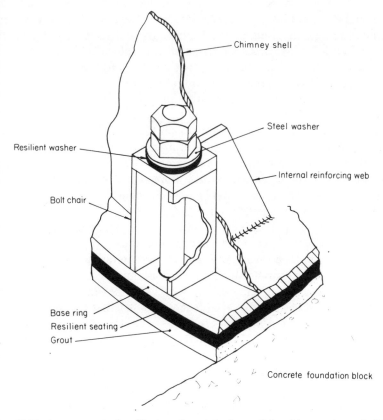

FIG. 12.21. Arrangement of resilient seating at the base of the chimney shown in Fig.
12.20.

12.4   PRINCIPAL FUNCTION OF IMPACT ABSORPTION

Although it is difficult to distinguish whether some applications are really vibration or impact applications, it is convenient to identify certain uses of resilient materials where energy absorption at impact is the prime function. One such application is to provide a resilient layer to cushion the impact between the steel anvil and the concrete inertia block of a forging hammer. In addition to resiliently supporting the foundation block to limit the transmission of ground-borne vibration from the hammer, a resilient interface is also needed between the anvil and the hammer to prevent localised crushing and even complete fracture of the reinforced concrete foundation block under the continuous pounding it receives during the hammer's operation. Traditionally, baulks of timber have been used for this purpose, but splintering and crushing of the timbers gradually take place, and the anvil then has to be lifted and the timbers replaced (Fig. 12.22).

Modern anvil mats are of high-strength multi-layer construction with proofed fabric layers alternating with thin layers of polychloroprene. After impact, the instantaneous downwards velocity of the anvil is of the order of 0·5 m/s, and the static pressure under the anvil increases from about $0.3 \, MN/m^2$ to perhaps $4 \, MN/m^2$ instantaneous peak pressure, depending on the stiffness of the resilient interface. There is a limit to the maximum acceptable flexibility of the interface because the anvil must not bounce more than a maximum amount at each blow. Maximum anvil dynamic movements of the order of 1 mm are current British practice, but there is a trend towards using softer anvil mats in certain foreign countries [7], where movements up to about 3 mm have been found acceptable and have led to a quieter hammer and reduced wear and tear during operation. It is important, when these larger anvil movements are permitted, that the anvil mat provides a source of damping as well as of flexibility, because the anvil must 'settle' rapidly after each blow so that the operator has time to adjust the workpiece before the next blow falls. With too little damping, the vertical rebound of the anvil after impact may cause it to lift off the resilient mounting for a brief period after every blow. Debris may then work progressively under the anvil and its resilient mounting, destroying the hammer's alignment, eating into the resilient mounting, and damaging the support surface of the concrete inertia block.

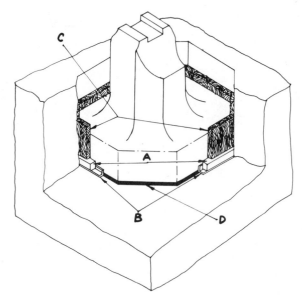

FIG. 12.22. Arrangement of a forging hammer anvil supported on a resilient mat instead of baulks of timber. A, Sealant; B, parting strip; C, timber packing; D, anvil pad.

Resilient buffer pads, generally of similar construction to hammer anvil pads, although sometimes with increased flexibility resulting from the inclusion of thicker layers of cellular filled materials, are extensively used in the mechanical engineering industry to cushion the impact of parts which may collide, and the rebound stops of a counter-blow hammer are a typical example (Fig. 12.23). Because these buffers are necessarily sited close to very hot components, they operate in a high-temperature environment, and it is important that their internal temperature does not approach a level at which early failure of the material might occur. In this case, damping in the buffer material has generally to be kept low to minimise the internal build-up of heat in the buffers during their repeated operation (Fig. 12.24).

A recent, unusual application of resilient buffers has been in connection with London Transport's development of a train arrester system to decelerate over-running trains safely at terminal stops (Fig. 12.25). London Transport's solution to this problem is to provide a bed of granite chippings at the end of the terminal line, and if a train

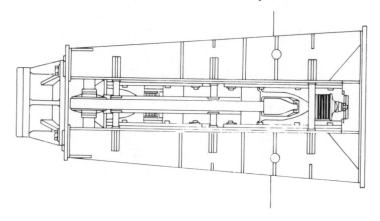

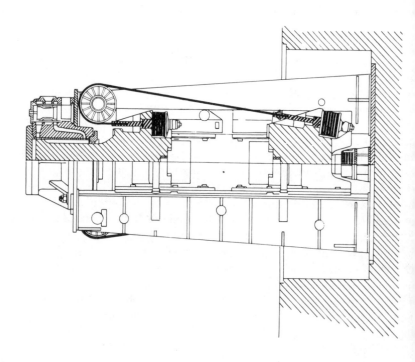

FIG. 12.24. A cork-modified rubber/fabric multi-layer buffer assembly for a counter-blow hammer.

ver-runs the station it is brought to rest gradually as it ploughs into 1e chippings. To prevent the train riding up over the bed of chip-ings, a steel sledge is provided which slides along the rails on PTFE kids, and displaces the granite chippings evenly in front of the train. ince this sledge itself weighs some 15 tonne, and it may be impacted y a train moving at about 30 km/hr, a resilient facing is needed on 1e front face of the sledge where it is impacted by the buffing plate f the train's leading motor car. This resilient buffer must reduce the everity of the initial impact sufficiently that the buffing strength of 1e impacting car is not exceeded and the front of the train is not

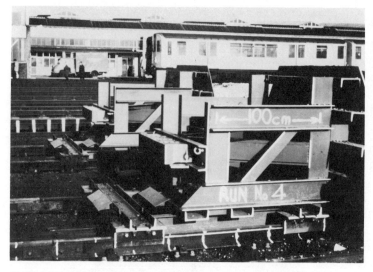

Fig. 12.25. LTE train arrester dolly facing assembly. (Courtesy of London Transpo
Executive.)

crushed before the gradual process of deceleration begins as th
sledge is pushed into the bed of chippings.

Special purpose elastomeric buffer assemblies have been develope
for this application, and their design is unusual for a number o
reasons. The frontal area of the buffer has to be small because of th
restricted area of the train's own buffer plate, the working stroke ha
to be large in relation to the overall depth of the buffer, and th
instantaneous peak pressure in the buffer material is very high. A
resilient material capable of accepting such large deflections an
pressure was therefore required, and it had to be incorporated in
buffer design which would permit the high lateral expansion occurrin
at impact while maintaining the overall stability of the whole assem
bly. The solution adopted was to build a multi-layer assembly wit
blocks of cellular-filled high-strength elastomer bonded to meta
spacers which slide over central guide bars. Full-scale crash tests o
this assembly have been successfully completed (Fig. 12.26) and
high-speed film of the deformation of the buffers at impact has bee
made to verify their dynamic performance. The peak dynamic forc
transmitted by the buffer is 120 tonne when the compressive stress is o
the order of 14 MN/m^2 and the dynamic deflection about 100 mm.

FIG. 12.26. Close-up of dolly facing after experimental impact. (Courtesy of London Transport Executive.)

Research into a new application to safely decelerate falling lifts is now in progress. This problem, which is a particularly important one for the mining industry, is to design a buffer which can withstand repeated low-velocity impacts without damage, but will progressively crush under a high-velocity impact to provide gradual deceleration of an over-running or falling lift cage. One solution to the problem which is now being evaluated consists of crushable metal cylinders, designed to deform plastically when a given peak load is exceeded, with a resilient cushion of a cellular filled elastomer which will not crush, but which is sufficiently flexible to safely attenuate low-velocity impacts without damage to the structure or injury to the occupants of the lift.

12.5 FLEXIBLE BELLOWS AND DUCT CONNECTORS

In recent years there has been an increasing interest in resilient bellows assemblies for the chemical-process industries and for furnace-gas exhaust ducting, gas-turbine exhaust systems, low-pressure steam pipework and air-conditioning ducts (Figs. 12.27–12.29).

The requirements are generally those where large deflections must

FIG. 12.27. Railway traction motor cooling bellows fabricated from metal-reinforce asbestos cloth, proofed with synthetic rubber.

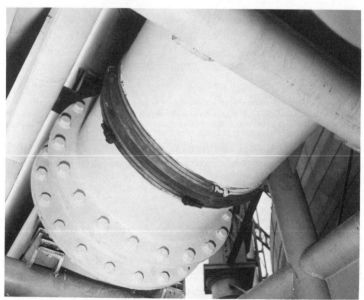

FIG. 12.28. Gas-turbine exhaust duct incorporating an expansion bellows of wove stainless steel/asbestos mesh, proofed with synthetic rubber.

FIG. 12.29. Composite exhaust bellows on a test rig at the National Gas Turbine Establishment. The bellows is fabricated from woven glass cloth, proofed with fluoro-carbon rubber, and incorporates a heat shield of bulk ceramic fibre.

be accommodated, or where high temperatures exist which formerly would have been catered for by bellows of a woven asbestos construction. A great deal of specialist development has taken place to produce elastomeric-based bellows to meet these various requirements and good results have been achieved under a variety of different operating conditions.

Low-temperature (below 150°C) applications have been satisfied by single half-convolution connectors formed to shape in a laminated EP rubber with nylon mesh reinforcement. By correctly proportioning the connector's shape, it is possible to accept ±25 mm lateral deflection of the two ends of a connector of total fitted length 50 mm, which is much greater than could normally be achieved with a stainless-steel unit for the same application.

For high-temperature applications, insulation must be provided to limit the operating temperature of the elastomeric-based materials, and a composite structure has been used with woven glass fabric or

218   D. E. Newland and A. D. Liquorish

ceramic fibre protecting the rubber-proofed outer skin of the bellows
One method is to locate the insulation inside a stainless steel o
Incoloy woven mesh and a PTFE skin may be included when operat
ing conditions are sufficiently acidic.

The main limitation at present is the maximum operating pressur
that can be accommodated by elastomeric bellows assemblies, but th
field is developing rapidly, and as materials and manufacturin,
methods improve the range of these devices will undoubtedly in
crease. A particular advantage of the units is their ease of trans
portation, as very large connectors can be assembled on-site, an
transported there in rolled or other packed form. This is importan
where large gas exhaust ducts have resilient connections which ma
be up to 100 m² in flow area.

REFERENCES

1. D. E. NEWLAND. 'Methods of dynamically testing rubber buildin
   materials under high static pressure'. In: *Proc. Soc. Envir. Eng. Sym
   posium on Dynamic Testing.* London, 1971.
2. D. E. NEWLAND and A. D. LIQUORISH. 'Progress in the development an
   testing of resilient building materials'. In: *Proc. Soc. Envir. Eng. Con
   ference on Vibrations in Environmental Engineering,* London, 197:
   Reprinted in *Noise Contr. & Vib. Reduction,* 1974, **5**, 4, 151.
3. D. E. NEWLAND and G. C. MCCREESH. 'Impact testing resilient materi
   als'. *J. Soc. Envir. Eng.,* 1976, **15**, 4, 11.
4. A. D. LIQUORISH. 'Structural resilient seatings'. In: *Soc. Envir. En
   Symposium on Developments in Resilient Mounting of Civil Engineeri
   Structures,* London, 1976.
5. JAMES WALKER & COMPANY LTD. (1975) *Walkers Crane Rail Strip—
   Technical Information.* Woking, Surrey: James Walker & Co. Ltd.
6. D. E. NEWLAND. *Proceedings of the Workshop on Railway and Tracke
   Transit System Noise,* Derby, England, 30 March–1 April 1976: Rap
   porteur's Report, Session **6**: 'Noise in elevated structures; vibratio
   isolation techniques'. *J. Sound Vib.,* 1977, **51**, 3, 449.
7. R. CIESIELSKI. *Description of an Experimental Investigation carried ou
   during the Rebuilding of the Foundation of an MPM 20 ton Hammer* (i
   Polish). Badania Doswiadczalne Konstruckcji, Krakow, Janowice, 1977

Chapter 13

RESILIENT MOUNTINGS OF LARGE STRUCTURES

P. GROOTENHUIS

13.1 INTRODUCTION

The improvement in the design and manufacture of elastomeric bearings to carry large loads in compression has a favourable influence upon the living and working conditions in large cities. Sites which were hitherto undesirable because of nearby surface railways and motorways, or with a network of underground train tunnels traversing the area, can now be developed for high grade occupation and for public usage without the intrusion of noise and vibration. The social implications of such developments are far reaching for life in the inner city. The requirement to isolate an entire building from ground-borne vibrations arises all the more frequently because of two separate and contrasting trends:

1) A demand for a better living and working environment with a minimum of disturbance from outside sources. The noise from road and surface rail traffic can be excluded by means of double windows and the installation of mechanical ventilation. However, the noise from an underground train—a low-pitched rumble—can then be heard all the more easily within a building with foundations close to a tunnel. Such a sudden intrusion from a hidden source can be most disturbing. The vibrations can often be detected and secondary rattling noises can contribute to the irritation.

2) Modern trends in the construction of buildings are towards lightweight floors and walls with the minimum number of joints. Such a construction is often 'dynamically lively', that is, once some vibration has been allowed to enter into the structure it can be transmitted readily to every part. The amount of vibrational energy dissipation in

high class forms of construction is rather small and almost negligib‖ for pre-stressed concrete and for a welded steel framework. A co◄ dition of resonance of a floor slab with a prominent input frequenc▮ can give rise to an amplification in vibration level of some 10 to 2▮ times.

The contrast in these two trends—a growing reluctance to accep▮ any intrusion within a building from say, a nearby railway, and th▮ readiness with which modern structures can transmit vibration—ca▮ present the developer and the architect with some difficult decision▮ However, when the decision is taken at an early stage in the design t▮ vibration-isolate the building or parts of it, then the structure can b▮ adapted to include resilient mountings and the additional cost nee▮ not be at all great.

The recommended background noise levels for different classes c▮ use of buildings, such as concert halls, hospitals, flats and offices ar▮ now generally accepted for design purposes [1]. A value for th▮ threshold of human sensitivity to vibration is not so well establishe▮ One of the lowest criteria has been set by Dieckmann [2] as ▮ vibration rating or K-value depending upon the amplitude (mean t▮ peak) and the frequency. A different rating is required for a perso▮ subjected to vertical and to horizontal vibrations. The now generall▮ accepted K-values are given in Table 13.1.

It can be seen that our perception of vibration is indeed very goo▮ e.g. at frequencies above 40 Hz in the vertical direction the amplitud▮ need be only slightly greater than 5×10^{-4} mm for the vibration to b▮ detectable.

TABLE 13.1
The Dieckmann K-values

Vertical vibration	Horizontal vibration
Up to 5 Hz, $K = Af^2$	Up to 2 Hz, $K = 2 Af^2$
5–40 Hz, $K = 5 Af$	2–25 Hz, $K = 4 Af$
Above 40 Hz, $K = 200 A$	Above 25 Hz, $K = 100 A$

A = amplitude in mm, f = frequency in Hz.
The regions for vibration sensitivity are defined as follows: $K = 0.1$, lower limit of perception; $K = 1$, allowable in industry for any period of time; $K = 10$, allowable for short durations only; $K = 100$, upper limit of strain for the average man.

The first complete building to be mounted on rubber bearings is
Albany Court, a block of flats which was built in 1964–65 over the
District Line railway at St James's Park Station in London. A detailed
description has been given by Waller [3]. One of the bearings is
shown in Fig. 13.1. The bearing was made from natural rubber with a
number of steel-reinforcing plates so that bulging under a compressive load can occur only in between the plates. A total of 13
bearings support the building, each located in a pocket in the foundation with a small clearance between the suspended part and the
rigid base so that a firm support will come into operation in the event
of a bearing failure. A close watch has been kept on the permanent
deflection and creep rate but this is remarkably low [4], slightly less
than 1 mm in seven years after the full compression had been taken
up during the first year. The occupants of these flats are unaware of
the many trains running immediately below the building.

Another early application of the suspension of a large structure on
rubber bearings is to be found in London at the Barbican Rede-
velopment where a section of two sets of double-track railway lines
have been mounted on a bridge deck on bearings to give a suspension
system with a natural frequency of about 7 Hz—for a detailed description see Grootenhuis [5]. It is obviously better and can well be

FIG. 13.1. A multi-laminate natural rubber bearing under Albany Court, London.

more economic to isolate the source of the ground vibration rather
than isolate buildings close to the disturbance. This is possible when
new railways are to be built or realigned. A floating track slab can be
introduced into a bored tunnel without necessarily an increase in
diameter [6]. Trains have been running over the Barbican suspended
track since 1965 and the occupants of the flats over the railway are
oblivious of this. Since these early installations many different types
of buildings have been isolated. To mention a few: a hospital [7], a
large council housing development [8], an office tower block [9] and
even a supermarket [10]. Many more examples could be quoted, e.g.
Davey and Payne [11]. It is not always necessary to suspend the
entire structure, only the highly sensitive areas need be isolated such
as a recording or broadcasting studio which would be constructed as a
suspended 'box-within-a-box' [12]. This form of construction will give
protection from disturbances from outside source and can isolate
one studio from another, which is often a much more difficult task. A
plant room can be a source of noise and vibration within a building
and it is sometimes advisable to provide a floating floor mounted on
rubber bearings [13]. It is, nevertheless, a specialised technique and
some guidelines with a warning have been issued in 1975 by the
British Standards Institution as a draft for development [14]. The very
recent application of supporting buildings on rubber bearings which
are very flexible in shear for protection against earthquakes will not
be described here (see Derham et al. [15]).

13 2   DESIGN OF RESILIENT MOUNTINGS

The response of a part of a building into which vibrations have
entered through the foundations will depend upon several local con
ditions. There can be a magnification of the input level of vibration
because of a condition of resonance, e.g. of a floor slab or of a
partition. On the other hand, a part of a building can be very
unreceptive to vibration because of a high local impedance, such as a
concentration of mass, large in comparison with other parts. For a
more detailed discussion of the transmission of noise and vibration
through structures reference should be made to Cremer and Heckl
[16], and Grootenhuis and Allaway [17]. The resilient mountings can
be placed either as part of the foundations, in groups of pads on the

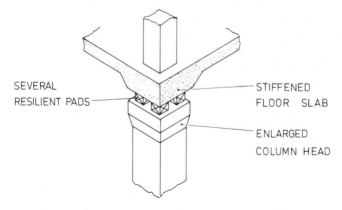

FIG. 13.2. Resilient mounting at elevated level.

pile caps and spread out on ground beams, or higher up the structure under walls and under floors on top of columns.

Figure 13.2 shows a resilient mounting for the upper part of a building. A number of pads are placed on the enlarged column head and the floor slab is then cast *in situ* on top, or alternatively, pre-cast planks can be used. Several resilient pads are normally used to make up a mounting rather than a single large pad. In this way different column loads can be accommodated using the same size of pads but varying the number per mounting. Keeping to the same compressive stress and the same pad dimensions with the same shape factor* will ensure a uniform static deflection and creep rate for all the mountings. It will also facilitate the provision of fail-safe upstands in between the pads and on the column centre line. Four types of resilient pad construction have been described in the B.S.DD 47 [14]:

(i) *Plain elastomeric pad.* A plain block of material which is normally suitable only for very low loads or for attenuation of the high-frequency vibrations with a thin pad. This type is not recommended for the mounting of large structures.

(ii) *Sandwich pad.* A plain elastomeric pad is bonded between two parallel metal plates which become the bearing surfaces. The plates prevent the lateral spreading and a greater load can be carried.

---

*Shape factor = cross-sectional area/area free to bulge.

(iii) *Multiple sandwich pad.* An even greater load carrying capacity can be obtained by controlling the general bulging of a sandwich pad with more metal reinforcing plates in between the bearing surface plates. The effective shape factor is then based upon the thickness of the elastomer in between the plates.

(iv) *Composite pad.* A pad made either from alternate layers of an elastomer and fabric reinforcement or of an elastomer with granular fillers and the fabric reinforcement. A wide range of properties can be obtained with this form of construction. Metal bearing plates can be bonded to these pads if necessary.

The exposed metal surfaces are normally protected against corrosion by cocooning with the elastomeric material. Protection against damage by fire has to be provided for mountings at elevated level in a building. There is no need for this when mountings are buried in the foundations. An additional layer of some 60 mm all round of natural rubber can provide such protection [18] by virtue of the low thermal conductivity. The detail design of the structures above and below the pads is of great importance as levelling, rigidity and structural integrity are all essential for the proper functioning of the mounting. Pads should be loaded centrally and uniformly, otherwise there is the danger of instability. Some lateral shear force can usually be accommodated and can thus resist all the wind loading unless the building is situated at a very exposed position or is very tall. Extra horizontal constraints will then have to be provided. The data required from the pad manufacturer for the design of a resilient mounting system, the test specification and precautions on site are outlined in the B.S.DD 47 [14].

A method for constructing the type of mounting shown in Fig. 13.2 at the top of a column and with the floor slab to be cast *in situ* is given in Fig. 13.3. The enlarged column head is to be made level by pouring on an epoxy grout and the bearings or pads are then bonded to it with a fail-safe upstand placed in between. It is essential that each bearing will have adequate space around it to bulge freely when the full load is applied. Great care must therefore be taken to prevent the ingress of wet concrete whilst casting the floor slab. One method is to fill the spaces in between the bearings with sand, place a polythene sheet over the mounting and cast the slab. The sand can be blown or washed out after the shuttering has been removed. Alternatively, the mounting can be covered with a steel sheet, preferably

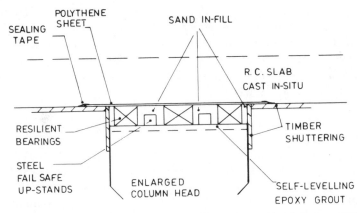

FIG. 13.3. A method of construction of a mounting at elevated level. R.C. = reinforced concrete.

galvanised which will be cast into the floor slab. There are several other variations for this method of construction.

13.3   DYNAMIC CONSIDERATIONS

The dynamic response of a structure with a resilient mounting, for example as described already in Fig. 13.2, will depend upon many 'local' conditions and upon the loading on the columns. The vibration levels on the floor slab, which are very important for setting the noise and vibration environment in the area, will depend upon how close the resonant frequencies are to the input frequencies, upon the amount of damping in the structure and upon the substructures carried on the slab. It has been shown by Fahy and Westcott [19] that the floor frequencies and damping can be affected markedly for long spans by the substructures, by finishes and by furnishings. A detailed analysis has to be carried out when the lowest floor natural frequency is approximately equal to the resilient mounting frequency. However, for most medium to short span floors the lowest frequency will be greater than the mounting frequency, hence for a first and conservative approximation the mounting frequency can then be obtained from

$$\text{mounting frequency (Hz)} = 1/2\pi\sqrt{kg/P} \qquad (13.1)$$

where k = dynamic stiffness (N/m); P = load (N); and g =gravitational constant ($9 \cdot 81$ m/s^2). A design value of between 7–12 Hz is often specified in order to provide some attenuation for the low end of the frequency spectrum of the input. The vibration transmissibility through a resilient mounting, in one direction and neglecting motions in other directions, is given by

$$\text{transmissibility} = \left\{ \frac{1 + \eta^2}{(1 - p^2/\omega^2)^2 + \eta^2} \right\}^{1/2} \tag{13.2}$$

where p = input frequency (Hz); ω = mounting frequency (Hz); and η = loss factor of the bearing material = $\tan \delta$. This expression is given in graphical form in Fig. 13.4, with the frequency ratio, p/ω, as the abscissa and for different values of the loss factor. A reduction in the vibration transmission can be obtained only for values greater than $\sqrt{2}$ of the frequency ratio. Below this value the motion can be

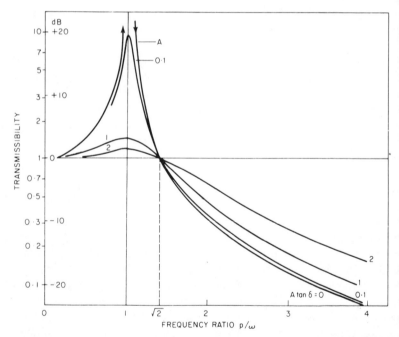

Fig. 13.4. Unidirectional transmissibility of a mounting as a function of frequency ratio for different values of the loss factor ($\tan \delta$). Curve A is for zero damping and the transmissibility goes to infinity at $p/\omega = 1$. The figure has been taken from Davey and Payne [11].

mplified and conditions can be worse than had no mounting been nstalled. Too much damping in the material is detrimental to effective vibration isolation at the higher frequencies. The 'roll off' of the ransmissibility curve at values of the frequency ratio much greater han $\sqrt{2}$ is approximately 12 dB/octave.

If follows from this simple analysis that with a mounting frequency of 7 Hz the transmission is reduced for input frequencies greater than 10 Hz and for a mounting frequency of 10 Hz only for frequencies greater than 14 Hz. Much of the vibrational energy in ground-borne vibrations from underground trains lies in the frequency range 20–150 Hz. The lower end of this frequency range is very close to the minimum frequencies at which attenuation can start. It is therefore crucial that the correct mounting frequency be specified and be achieved in practice. This requires an accurate estimate of the load P) on the mounting and a fairly precise knowledge of the *dynamic* stiffness of the actual bearings. The load can be obtained accurately from the detail design of the structure and from a realistic assessment of the live loading for the building. An increase in live load in practice will decrease the mounting frequency which is then beneficial. There is less certainty in obtaining an accurate estimate for the dynamic stiffness of the actual bearings to be used in a mounting. There is some confusion in the literature about the dynamic stiffness and the static deflection. Davey and Payne [11] have recognised the difference between the dynamic and static stiffness but, most unfortunately, they have given a curve of the natural frequency of a rubber spring as a function of the *static* deflection (their Fig. 8.1). Much of what follows is based upon this incorrect evaluation of the natural frequency. More recently Derham *et al.* [15] have given the frequency of a mounting for buildings in terms of the 'effective deflection', which is the static deflection reduced by a dynamic to static factor for the material from which the bearings are to be made. The actual dynamic stiffness cannot be determined in this manner for all kinds of elastomeric bearings. The dynamic to static stiffness ratio is fairly small for bearings made from natural rubber, usually between 1·5 and 3, but can be much greater for composite bearings. The ratio depends not only upon the material but also upon the bearing design, shape factor and the loading. The natural frequency should therefore not be determined from a measurement of the static deflections but by a dynamic resonance test with the nominal load applied.

One of the differences between the dynamic and static conditions is that the dynamic strains are so very much smaller than the overall static strain for a resilient bearing. It is well known that the load-deflection curve in compression for an elastomeric material is much steeper at small strains than the average for large strains and that the slope of a small incremental load–deflection variation superimposed upon an already applied large strain is greater than the slope of the main load–deflection curve at that point—this is known as Mullins effect. The dynamic stiffness can be even greater than these small strain static values. The dynamic stiffness is dependent upon the frequency of the vibration and sometimes upon the duration under load. Some measurements by Grootenhuis [20] on a composite bearing material (now withdrawn from the market) gave a 170% increase in the small strain static stiffness whilst under a compressive load for some 1000 hr as compared with the value obtained after the first hour under load. This increase continued the longer the test went on. Much has been written about the stress softening of elastomeric materials when subjected to large strains but why there should be an increase in small strain stiffness and in dynamic stiffness is not so fully understood. The vibration behaviour is linked possibly with the dynamic stress fields around the filler and other inclusions as has been determined for fibre reinforced viscoelastic materials by Paipetis and Grootenhuis [21].

13.4 EXAMPLES OF BUILDINGS ON SPRINGS

A description will be given of a few recent installations with which the author has been concerned.

13.4.1 Wellington Hospital

Wellington Hospital (Fig. 13.5) is a good example of a development where a hitherto undesirable site was the only one available. A developer wanted to build a private clinic in the centre of London and it had to be close to the medical consultants in Harley Street. The only economically viable site was an old cutting for one of the main railway lines out of Marylebone Station. The cutting lay just north of the Lord's Cricket Ground alongside Wellington Road between two sections of tunnels, and had been used as a working area when the tunnels were dug. It had remained vacant ever since. The sides of the cutting were massive retaining walls of brick and it looked at first

FIG. 13.5. The Wellington Hospital beside Wellington Road, London.

sight a very suitable location for a small hospital, the hole for a basement being there already [7]. But the diesel commuter trains still run through that cutting and would thus have to run right through the basement. In addition the tunnels for the Bakerloo and the Metropolitan Line underground trains are nearby under the busy Wellington Road, not a very desirable site. A cross-section of the site and of part of the building which has been erected is given in Fig. 13.6. A tunnel for the Metropolitan Line comes within $3\frac{1}{2}$ m of the new structure. Some of the operating theatres are on the ground floor, just one level above the roof of the box built for the trains from Marylebone Station. A diesel train can be seen in Fig. 13.7 passing through the box in the basement whilst under construction.

The ground vibrations at pile cap level from both the mainline and the Bakerloo Line trains were strong enough to be noticeable and the noise level was totally unacceptable. A heavy concrete box with a roof 1-m thick was constructed around the mainline track, as can be seen in Fig. 13.7. In addition an acoustic plenum chamber was provided at lower ground floor level (Fig. 13.6). The main part of the hospital has been protected from the ground- and structure-borne noise and vibration by isolating every column on natural rubber

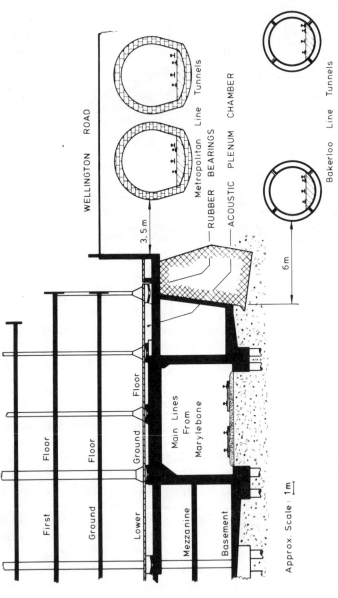

Fig. 13.6. Part cross-section of the Wellington Hospital with the box tunnel for the mainline trains from Marylebone and the two sets of underground train tunnels.

FIG. 13.7. The box tunnel under construction showing a train from Marylebone Station, London.

bearings to give a mounting frequency of about 7 Hz. Neither the patients nor the surgeons are aware of the trains passing underneath them, although the noise and vibration in the basement are quite alarming.

13.4.2 Alexandra road

The borough of Camden has embarked upon a major housing development along the busy mainline railway out of Euston Station, which carries intercity, goods and local trains. Part of this development is known as Alexandra Road [8]. In order to shield the greater part of the site from the noise of the trains, a large barrier block of flats has been erected very close to the railway. Only the minimum number of openings have been allowed in the facade on the site of the railway, and then only with widely spaced double windows. This barrier block is shown in Fig. 13.8 with an intercity train just visible passing behind it. The noise and vibration level inside this block would have been intolerable had the entire structure not been isolated.

It was decided to place the resilient mountings at pile cap level in the foundations, to avoid intersecting the large shear walls which provide structural integrity for the building. Placing the mountings in the covered-over foundations removed any danger of damage by fire. Figure 13.9 shows the resilient mounting. The resilient bearings or pads were bonded to the prepared base with an epoxy mortar. A pre-cast concrete beam was placed on top and a shear wall or set of columns erected on top. The backfill was kept away from direct contact by means of a resilient facing applied to the concrete. The mountings were sealed with a polythene sheet anchored above and below to the concrete and weighted down with sand. A detailed noise and vibration survey has not yet been carried out.

13.4.3   Rank Hovis McDougall office tower

This office tower is an example of a change in circumstances after the site had been acquired in that the new Victoria Line tube tunnels were planned to pass under it. The location is along the Thames close to Vauxhall Bridge, and because of the special measures which are necessary when boring a tunnel under a tidal river it was not thought feasible at that time to incorporate a floating track slab in that section. The tunnels were to be bored in between the piles and there was thus a very good transmission path for vibrations to enter into the building and an adjacent block of flats. The surrounding main roads create a severe traffic noise environment and double windows with mechanical ventilation were to be used. Vibration isolation of the entire building and the flats was specified [9]. The office tower is shown in Fig. 13.10. The tower has 20 floors and a height of 66 m above ground. The rigidity of the tower is provided by a central core of lift shafts and staircases, and the wings are supported on columns. The dynamics of sway of the tower during strong winds became the controlling factor in the design of the resilient mountings rather than the mounting

FIG. 13.8. The barrier block at Alexandra Road beside the main lines from Euston, London.

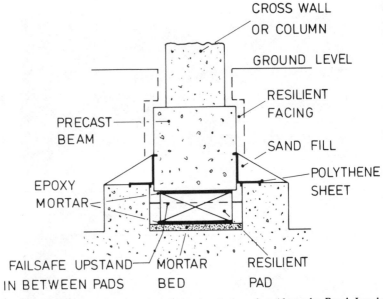

FIG. 13.9. Resilient mounting at foundation level as used at Alexandra Road, Londo

frequency for maximum vibration isolation. Fortunately, the locatio
is approximately midway between stations and the trains woul
normally operate at full speed. The input frequencies would therefor
be somewhat higher than at other locations. A mounting frequency c
slightly greater than 20 Hz was aimed at.

A view of four resilient pads to form a mounting on top of a pil
cap is shown in Fig. 13.11. Four stainless-steel dowl pins wer
provided to act as fail-safe sideway stabilisers to control any sever
horizontal movement or settlement. A rubber sleeve was placed ove
each dowl pin to prevent the transmission of vibration (see Fi
13.11). The steel reinforcing bars with a steel plate and a mortar be
were than placed on top (Fig. 13.12). This is not the same column a
shown in Fig. 13.11. The dowl pins can be seen protruding upwards i
between the reinforcement. Shuttering was then constructed and
view of a completed column is shown in Fig. 13.13. A differe
construction was used for the core of the tower with a number c
pads under the walls and under the floor. The pads were bonded
the base slab in a pattern to ensure a deflection similar to th
experienced by the column mountings. Pre-cast reinforced concre

FIG. 13.10. The Rank Hovis McDougall office tower near Vauxhall Bridge, London.

units were then placed on the pads as shown in Fig. 13.14. The joints were then sealed and further concrete poured *in situ*. A row of pads can be seen on the side in Fig. 13.14 for a wall.

Simultaneous measurements have been taken of the vibration below and above a mounting on a column when a Victoria Line train passed underneath. A typical vibration trace is shown in Fig. 13.15 with the sensitivity for the trace of vibration above the mounting 1·35 times greater than for the trace of the input vibrations. It can be seen that the higher input frequencies have been much reduced and that only some low-frequency motion persists. The frequency of this is

FIG. 13.11. A column mounting with four resilient pads, dowl pins and rubber sleeves.

about the same as the mounting frequency and this is well below the audible range. Neither train noise nor vibration can be heard or detected in the office accommodation, although the trains can be heard in the car park below the mountings. The tower has withstood several severe gales without any apparent effects.

13.5 CONCLUSIONS

The technique of vibration isolating large structures on resilient mountings has been established for a number of years. This can lead to considerable improvements in inner cities by the development of hitherto undesirable sites. The elastomeric bearing and pad materials now available give the designer a wide scope in the layout of the mountings but it is best to consult the specialists. It is essential to know the dynamic properties of the bearing and pad designs, and more work is required to investigate the variation in the difference between the dynamic and static stiffnesses.

ACKNOWLEDGEMENTS

he structural consulting engineers and the resilient bearing
nanufacturers for the three installations described were, in al-
habetical order, André Rubber Co. Ltd, R. J. Crocker & Partners,
nthony Hunt Associates, Ove Arup and Partners, and TICO–James
Valker & Co. Ltd.

FIG. 13.12. Reinforcing steel bars placed in position on top of the pads.

FIG. 13.13. A completed column mounting, except for finishes.

FIG. 13.14. Resilient pads in position ready to receive walls and pre-cast floor unit

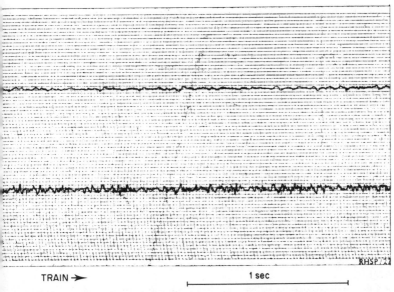

TRAIN →   1 sec

IG. 13.15. Simultaneous vibration recordings above and below a column mounting at ie Rank Hovis McDougall tower. Top trace: above pads, sensitivity 1·35. Bottom trace: below pads, sensitivity 1.

REFERENCES

1. Anon.(1970). *Guide to Current Practice*, Vol. A. section A-1, Chartered Institute Building Services.
2. D. DIECKMANN. 'A survey of the influence of vibration on man'. *Ergonomics*, 1958, **1**, 347.
3. R. A. WALLER. (1969). *Building on Springs*. Pergamon.
4. Anon. 'Long term tests confirm laboratory predictions'. *Rubb. Dev.*, 1975, **28**, 7.
5. P. GROOTENHUIS. (1967). 'The attenuation of noise and vibration from railways'. In: *Proceedings of Conference on Environmental and Human Factors in Engineering.* University of Southampton, 1967.
6. P. GROOTENHUIS. 'Floating track slab isolation for railways', *J. Sound Vib.*, 1977, **51**, 443.
7. Anon. 'Luxury hospital'. *The Consulting Engineer*, 1974, **38**, S 25.
8. Anon. 'Housing at Alexandra Road'. *The Architectural Journal*, 1976 (September), 441. See also *The Consulting Engineer*, 1977, **41**, 10.
9. R. J. CROCKER. 'Construction of the RHM Centre on resilient mountings'. *Insulation*, 1972, (July), 173.
0. J. SOWRY. 'Supermarket sold on sound and vibration isolation techniques'. *J. Noise Control & Vibration Reduction*, 1974 (November), 327.

11. A. B. DAVEY and A. R. PAYNE. (1964). *Rubber in Engineering Practic* London: Maclaren.
12. A. BURD. 'Structural implications of good sound insulation'. *J. Env. Eng* 1977, **16**(3), 27.
13. P. H. ALLAWAY. 'Floating floors–an appraisal', *J. Env. Eng.*, 1977, **16**(2 3.
14. British.Standards Institution, DD 47. (1975). *Vibration Isolation of Stru tures by Elastomeric Mountings.*
15. C. J. DERHAM, L. R. WOOTTON and S. B. B. LEAROYD. 'Vibratio isolation and earthquake protection of buildings by natural rubbe mountings'. *N. R. Tech.*, 1975, **6**(2), 21.
16. L. CREMER and M. HECKL. (1972). *Structure Borne Sound.* Translate and revised by E. E. UNGAR. Springer-Verlag.
17. P. GROOTENHUIS and P. H. ALLAWAY. (1971). 'Noise and vibratio nuisance within buildings'. In: *Proceedings of Symposium on Enviro mental Aspects of Pollution*, London. Soc. Env. Eng.
18. C. J. DERHAM and A. P. PLUNKETT. 'Fire resistance of steel laminate natural rubber bearings'. *N. R. Tech.*, 1976, **7**(2).
19. F. J. FAHY and M. E. WESTCOTT. 'Measurement of floor mobility at lo frequencies in some buildings with long floor spans'. 1978, *J. Sound Vib* **57**(1), 101.
20. P. GROOTENHUIS. 'The anti-shock mounting of testing machines', *Pro Instn. Mech. Eng.*, 1965, **180**(part 3A), 388.
21. S. A. PAIPETIS and P. GROOTENHUIS. 'The dynamic properties fibre-reinforced viscoelastic composites', to be published.

Chapter 14

REINFORCEMENT OF ELASTOMERS BY SILICA*

A. VOET, J. C. MORAWSKI and J. B. DONNET

14.1 INTRODUCTION

During the past decades important progress had been made in the understanding of reinforcement of elastomers by carbon blacks [1], but no comparable advances had been achieved for reinforcement by silica. The following investigation is an attempt to elucidate the role of silica as a reinforcing agent in elastomers, adapting in part methods proven successful for carbon blacks and developing in part methods more suitable to the specific problems of silica.

The general approach used in this study was to prepare a number of silica pigments of widely varying characteristics and morphology and to incorporate these products, as well as a number of commercially available silicas, into an elastomer. The determination of a large number of physical-chemical properties, by conventional and by specially designed methods, for both pigments and vulcanisates, permitted us to establish certain relations between selected properties of the silicas and of the vulcanisates by regression analysis.

14.2 EXPERIMENTAL

14.2.1 Silica

Besides eight commercial silica pigments, we included twelve specially prepared silicas. The pigments used had surface areas, as

---

*This paper was presented at the symposium by Dr. A. Vidal and originally appeared in *Rubber Chem. Technol.*, 1977, **50**(2), 342.

measured by low temperature nitrogen adsorption, of 60–410 m²/g external surface areas, as found by CTAB adsorption [2], of 40 250 m²/g; particle diameters of 13–40 nm and DBP absorption num bers of 70–380 cm³/100 g. Moreover, silicas were prepared with pH (o a slurry of 5 g in 100 cm³ water) of 3·5–9·0, without much variation i other properties.

Besides these properties, the following characteristics of the silica were determined: external surface areas by means of low-temperatur nitrogen adsorption by the Lippens–de Boer t method [3], bot without ultramicropores (diameters below 0·5 nm) as well as withou pores of diameters above 2 nm; specific volume at a pressure o 60 MPa and the change of specific volume with pressure, which ar both structural parameters for carbon blacks.

14.2.2 Vulcanisates

Vulcanisates used in the regression analysis were prepared accordin to the following recipe: SBR 1509 100; silica 50; stearic acid 5; PE(4000 (polyethylene glycol MW 4000) 3·0; sulphur 2·3; MBT: (dibenzthiazyl disulphide) 1·2; Permanax CD (4,4'-bis(phenylisoprop) lidene) diphenylamine) 2·0; Rhodifax 16 (N-cyclohexyl-2-benzthiazy sulphenamide) 1·2; DOTG (N,N'-di-o-tolylguanidine) 1·4; TMT) (tetramethylthiuram disulphide) 0·2.

Other recipes used will be given where applied. The vulcanisate were tested for the conventional static mechanical properties, such a stress at 100 and 300%, tensile strength, elongation, Shore hardnes: rebound, tear strength (crescent and die), permanent set, and he; buildup. Furthermore, extensive dynamic mechanical measuremen were made, such as dynamic moduli and loss angles at differer amplitudes, frequencies, and temperatures, at different static elor gations [4]. Measurements analysed here were done at 30°C, 11 Hz, a a double strain amplitude of 0·16% for static extensions of 0–50%. I addition, samples were tested by Monsanto rheometer, determinin maximum and minimum torques. Additional properties measure were the adhesion between silica and elastomer, by swelling i solvents, by volume change on deformation, and by electron micrc graphy, both by scanning and transmission methods. Finally, a stud was made of the degree of dispersion of the silica in the elastomer, b optical and electron micrography and by radiography.

14.3 RESULTS

4.3.1 Correlation between silica and vulcanisate properties

he experimental data were treated by a computer program to obtain
ıe coefficients a_i in a multilinear regression equation

$$\bar{Y} = \sum_{i=1}^{n} a_i \bar{x}_i + b, \qquad (14.1)$$

here $\bar{Y}$ is the dependent variable and $\bar{x}_i$ the independent. The
ılculations were carried out by the method of the least squares.
fter determining the correlation matrix, the program introduces the
dependent variables in the regression one by one. In the first step,
ıe variable $\bar{x}_i$ is introduced which has the strongest correlation with
. At each consecutive step, the independent variable is introduced
hich has the best correlation with the formulation of the previous
ep. Table 14.1 indicates some of the results. Some of the above
:sults are represented graphically in Figs. 14.1–14.3.

4.3.2 Structural index

has been shown that a linear relation exists for carbon blacks
etween specific volume and logarithm of applied pressure in the
ınge of about 1–70 MPa. For most furnace blacks the lines can be
xtrapolated to a single, common point [5]. The slope of these lines,
; well as the value of the specific volume at any given pressure in the
ınge, appears to be indicative of the carbon black structure.
  We have adapted this approach to silica and measured specific
olumes at pressures varying from 1 to 70 MPa. A typical graph, as
ıdicated in Fig. 14.4, shows the relation for three silicas made from
ıe same press cake with three different drying treatments. It may be
:en that there are three different linear relations at pressures be-
veen 1 and 12 MPa, which merge and form a single line at 12 to
) MPa. Thus, it is apparent that the slope at the higher pressures is
ıdependent of the treatment of the filter cake, as has been noticed
ır a number of different silicas. Contrary to the observation for
ırnace blacks, however, there is no common point of these lines for
ifferent silicas. Apparently the following equation holds,

$$V_{sp} = I_S \log P + b_S, \qquad (14.2)$$

TABLE 14.1

Correlations between certain properties of silica and its SBR vulcanisate

Silica	Vulcanisate[a]	Correlation coefficient[b]
External surface area (CTAB)	Crescent tear	0·735
	Heat buildup	0·946
	E'_0	0·833
	$E'_0 - E'_{50}$	0·836
	E''_5	0·778
	Rebound	−0·711
Structural index[c], I_S	ΔT (max.–min. torques)	0·834
	Tensile strength	0·864
	Stress at 300%	0·495
	Elongation	0·823
	Shore hardness	0·955
	Die tear	0·754
	E'_0	0·795
	E'_{50}	0·803
	E''_5	0·678
DBP absorption	Min. torque	0·723
	Tensile strength	0·794
	Rebound	−0·829
	Heat buildup	0·793
	$E'_0 - E'_{50}$	0·697

[a] E'_0 and E'_{50} refer to Young's dynamic elastic modulus at 0% and at 50% sta extensions, respectively, while E''_5 refers to the corresponding viscous modulus at static extension, all measured at 30°C and 11 Hz with an amplitude of 0·16%.
[b] For 19 pairs of observations there are 17 degrees of freedom. For correlati coefficients of 0·456 and 0·693, the probability threshold values are 95% and 99·9 respectively.
[c] Slope of specific volume log pressure relation at elevated pressures (see secti headed 'Structural Index').
CTAB = cetyl trimethyl ammonium bromide, DBP = dibutyl phthalate.

in which I_S, denoting structural index, is a coefficient characteristic each silica while b_S is a constant for the silica.

14.4 EVALUATION OF ADHERENCE BETWEEN SILICA AND ELASTOMER

14.4.1 By swelling in solvents

Kraus [6] has developed a theory for the swelling of filled elastome in solvents based on the assumption that the swelling is zero at t

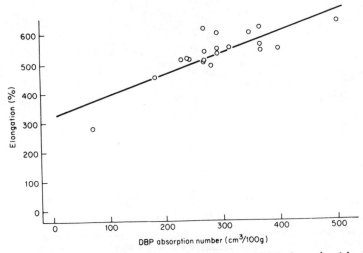

FIG. 14.1. Elongation at break (ordinate) as a function of DBP absorption (abscissa) (correlation coefficient 0·823).

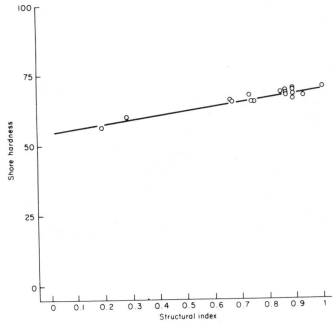

FIG. 14.2. Relation between structural index (abscissa) and Shore hardness (ordinate) (correlation coefficient 0·955).

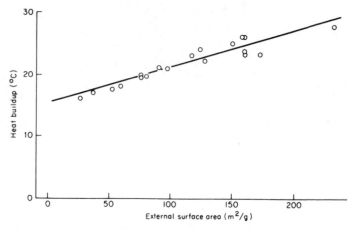

FIG. 14.3. Heat buildup (ordinate) as a function of external surface area (abscissa) (correlation coefficient 0·946).

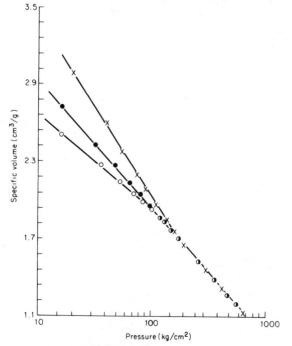

FIG. 14.4. Specific volume (ordinate) as a function of logarithm of pressure (abscissa) for three silicas: ×, quick dried; ◑, slowly dried; ○, slowly dried and milled ($10 \, kg/cm^2 = 1 \, MPa$).

particle surface and increases progressively at increasing distances from the interface until its normal value has been reached. Figure 14.5 shows the calculated relationship between V_{ro}/V_r and $\phi(1 - \phi)$ for particle–elastomer adherence and for nonadherence, in which V_r is the volume fraction of rubber in the gel of the filled vulcanisate after swelling, V_{ro} the volume fraction of the pure gum after swelling, and ϕ the fraction of the filler in the vulcanisate before swelling. Figure 14.6 shows the actual values we obtained for sulphur- and for peroxide-cured vulcanisates, in n-heptane and in benzene.

The sulphur-cured vulcanisates were made according to the following recipe: SBR 1509 100; ZnO 3; stearic acid 1; Rhodifax 16 1·2; DOTG 1·4; TMTD 0·2. Silica added was 0, 5, 25, 40, 50, 60, or 70 phr, while PEG 4000 was added in a quantity of 6% of the silica. Peroxide-cured vulcanisates were prepared according to the following recipe: SBR 1509 100; Permanax 45 (polymerised trimethyl-2,2,4-dihydro-1,2-quinoline) 2; DiCup 40 C (40% active dicumyl peroxide) 2·5; silica: 0, 5, 25, 40, 50, 60, or 70.

Figure 14.6 shows that the relations of Fig. 14.5, required by the theory of Kraus, appear to be valid up to about 40 phr of silica. The distinct failure of these relations at higher concentrations would seem to indicate the onset of dewetting and vacuole formation, no longer fulfilling the conditions set forth in the Kraus theory. Clearly, at

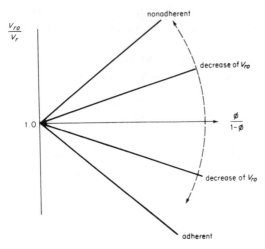

FIG. 14.5. Scheme of calculated swelling relations for adherent and nonadherent fillers in a vulcanisate. Ordinate: V_{ro}/V_r; abscissa: $\phi/(1 - \phi)$.

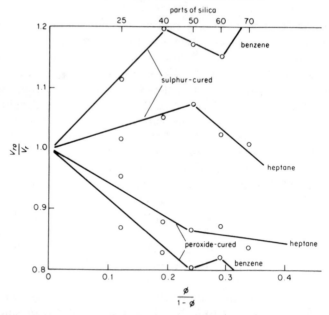

FIG. 14.6. Swelling as a function of degree of loading for sulphur- and for peroxide-cured vulcanisates in benzene and in heptane (coordinates as in Fig. 5).

loadings below the limiting pigment concentration, the curing system determines the slope of the lines. Peroxide-cured systems have negative slopes, pointing to adherence, while the sulphur-cured system studied has a positive slope, indicating nonadherence, between silica and elastomer.

Figure 14.7 shows the swelling relations for two different series of sulphur-cured vulcanisates (Table 14.2). The TMTD system leads to a

TABLE 14.2
Recipes used for sulphur-cured vulcanisates

No.	1	2	3	4	5	6	7	8
SBR 1509	100	100	100	100	100	100	100	100
Silica	—	20	40	60	—	20	40	60
Sulphur	2·5	2·5	2·5	2·5	2·5	2·5	2·5	2·5
TMTD	3·8	3·8	3·8	3·8	—	—	—	—
DMDC-Zn[a]	—	—	—	—	2·5	2·5	2·5	2·5

[a] Zinc dimethyldithiocarbamate.

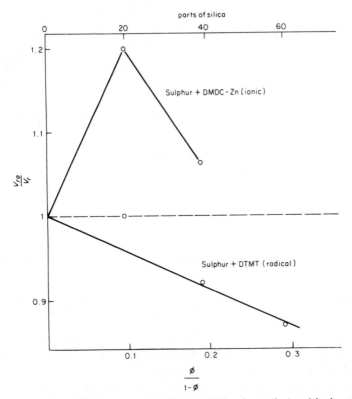

FIG. 14.7. Swelling in benzene as a function of loading for radical and ionic sulphur-cured SBR-silica vulcanisates (coordinates as in Fig. 5).

radical vulcanisation mechanism [7], while the DMDC-Zn system results in an ionic mechanism [8, 9]. The data of Fig. 14.7 indicate that the radical mechanism leads to strong particle–elastomer adherence. Ionic vulcanisation has a more complex character: nonadherent at lower filler concentrations, becoming weakly adherent at higher concentrations. Figure 14.8 shows the strong influence of the pH of the silica on swelling for a loading of 50 phr in SBR. It is remarkable that the degree of swelling is practically independent of the pH at a loading of 40 phr.

14.4.2 By volume change on deformation

We measured the change in volume with deformation of SBR vulcanisates reinforced with silica, in the presence and absence of a

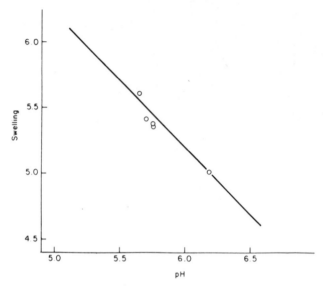

FIG. 14.8. Influence of pH (abscissa) of silica on swelling in benzene (ordinate) of 50 phr silica–SBR vulcanisates (correlation coefficient 0·969).

mercaptosilane coupling agent. The recipe used was the following: SBR 1509 100; silica 50; PEG 4000 3; stearic acid 3; ZnO 3; MBTS 0·75; DOTG 1·5; sulphur 2·2; Permanax 49 HV 2; hydrocarbon oil 6. To one of the two samples 1·0 phr of mercaptosilane had been added.

Samples were submerged in water in a dilatometer and were stretched by means of a wire which was wound inside the instrument. The liquid level in the capillary was measured by means of a cathetometer.

A typical result is shown in Fig. 14.9, where the volume change has been indicated as a function of the tensile stress at various extensions. It is clear that, in the presence of the coupling agent, the volume change of the vulcanisate is markedly smaller when strained to the same stress than in the absence of the coupling agent.

14.4.3 By scanning electron micrography

Figure 14.10 shows a dewetted grain of silica. Figure 14.11 indicates that the vacuole created by the dewetting process may lead to the formation of microtears, thus materially reducing tensile and tear strength of the vulcanisate.

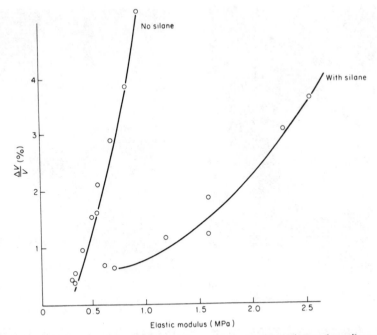

FIG. 14.9. Relations between change in volume (ordinate, $100 \, \Delta V/V$) and tensile stress (abscissa, MPa) for silica–SBR vulcanisates. Silica untreated and treated with silane.

14.5 DISPERSION OF SILICA

Measurement of the degree of dispersion of a silica in an elastomeric matrix was carried out by optical and electron microscopy as well as by radiography. Optical microscopy, difficult, if not impossible with silica, in view of the nearness of the refractive indices of filler and matrix, is materially aided by adsorption of a dye. Thus, by swelling a vulcanisate section about 20 μm thick in a benzene solution of methyl red, the dye is adsorbed on the particle surface, thus increasing the optical contrast. After solvent removal, the sample is immersed in a liquid of the same refractive index as the elastomer to eliminate the effect of surface irregularities. Observations by optical microscopy as well as by radiography reveal the presence of oversize grains under conventional conditions of dispersion.

A more precise image of the state of dispersion of the silica in the

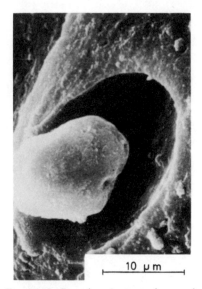

FIG. 14.10. Scanning electron micrograph of dewetted grain of silica showing vacuole.

FIG. 14.11. Scanning electron micrograph of microtear initiated by elimination of dewetted grain of silica.

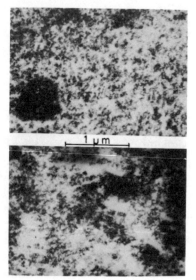

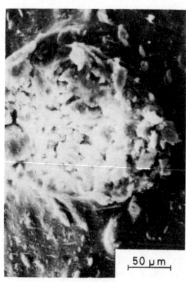

FIG. 14.12. Transmission electron micrograph of SBR-silica vulcanisate with (A, top) and without (B, bottom) mercaptosilane.

FIG. 14.13. Scanning electron micrograph of particle clump of incompletely wetted silica.

astomer was obtained by transmission electron micrography of
:ctions of the vulcanisate about 0·1 μm thick. Figure 14.12 shows
ιe results for the silica dispersions of Fig. 14.9. We observe the
resence of agglomerated silica aggregates (here in black) forming a
ɔntinuous network. It is also clear that mercaptosilane leads to
etter dispersion (compare Fig. 14.12A and B). It would seem that the
nprovement in mechanical properties in the presence of mercap-
ɔsilane is not only caused by improved pigment–elastomer adherence
ut also by a better dispersion.

Scanning electron micrography reveals that, even under the best
ommercial conditions of dispersion in a Banbury, a number of larger
lumps of silica, of diameters of 10–100 μm, are found in vul-
anisates. Moreover, many of these clumps are not properly wetted,
s shown for example, in Fig. 14.13. It is quite obvious that the
resence of such unwetted grains of silica will cause a drop in the
ɩechanical strength characteristics of the vulcanisate.

14.6 DISCUSSION

Ve found that three characteristics of the silica pigments correlate
vell with the static and dynamic mechanical properties of silica-
einforced SBR vulcanisates. The external surface area, as measured
·y CTAB adsorption, appears to be an important parameter. This is
ιot the case with the surface area as measured by low-temperature
ιitrogen adsorption, with or without pore areas as measured by the
-method of Lippens–de Boer. This observation seems reasonable
ince only the external surface interacts with the large elastomeric
nolecules. Apparently the interaction of CTAB with silica follows a
imilar pattern to elastomer–silica interaction, different from nitrogen
ιdsorption. Pore corrections are not satisfactory for silica, in contrast
vith the case of carbon blacks [2].

The second important parameter is the structure as indicated by the
ƆBP absorption number. The effect of structure of a reinforcing filler
ɔn reinforcement is well known, and the observation is not un-
:xpected. The fact, however, that the structural index appears to be
ɩn important parameter is a surprise.

In order to understand these findings, we must bear in mind that the
ƆBP absorption number is, in fact, a reflection of the total structure,
ɩ combination of persistent (primary) and transient (secondary)
;tructure, since the liquid demand of the pigment depends on the

space available between the elementary pigment particles, whether
that space has been created by elementary particles joined by chem
ical bonds or by Van der Waals attraction. The structural index, on the
other hand, appears to relate more to the persistent structure.

In the lower pressure ranges we observed a different structura
index than in the higher pressure ranges. Moreover, at the lowe
pressures it depends on the pretreatment of the silica, such as mode
of drying and grinding. It would seem that in the lower pressur
ranges the structural index reflects aggregation and agglomeratio
combined, while in the higher pressure ranges agglomeration is sup
pressed or even eliminated by the effect of the pressure, which force
particles to move into available spaces. Thus, the higher pressur
structural indexes would seem to reflect aggregation only.

Another interesting observation is that it is not the specific volume
itself at a high pressure which is an important silica parameter, bu
rather its change with pressure. This behaviour differs from that o
furnace blacks, where the specific volume versus log pressure line
converge to a common point, corresponding to closest random pack
ing, resulting in an equivalence of actual values and slopes. It would
seem that the furnace black aggregates are able to slide into a
close-packing configuration under pressure, while the silica aggregate
are not. The structural index, indicative of the resistance to inter
particle space elimination by pressure, apparently also marks the
extent of space retention under static and dynamic conditions in the
vulcanisates, where this structure is subjected to a variety of pres
sures. The importance of the interparticle space, the direct cause of
occlusion of the matrix is well known [10].

The study of the adherence between silica and SBR showed that in
contrast to reinforcing carbon blacks, the silica does not have perfec
adherence to SBR. The measurement of the volume change or
deformation, a direct measure of dewetting, is usually carried out at
fairly high extensions (up to 500%). Swelling tests in benzene
however, lead to a volume swelling of 320%, or, presuming an affine
swelling, to a linear deformation of 62%. In heptane, the linea
swelling is only 27%, yet dewetting is observed, indicating its occurrence at rather small deformations for loadings above about 45 phr.
Both swelling in solvents and volume change on deformation indicate
that the phenomenon of dewetting of silica greatly influences reinforcing properties for SBR.

The improvement obtained in mechanical properties as a result of e incorporation of a silane coupling agent in the vulcanisate onfirmed the well-known effect of better pigment–elastomer adserence, although part of the improvement must be due to enhanced ispersion.

The conclusions derived indirectly from swelling and volume inrease on deformation were confirmed by electron micrography, nowing actual vacuole formation.

A most interesting observation is the influence of the system of ulcanisation on silica–SBR adherence. Thus, peroxide-cured systems enerally lead to better adherence than sulphur-cured systems, while radical sulphur-curing system showed a much better pigmentlastomer adherence than an ionic sulphur-curing system. While, in einforcing carbon blacks, the pigment–elastomer interaction occurs redominantly on mixing alone [11], it would seem that with silica the ulcanisation step, and therefore the vulcanisation system, plays a ominant role.

The influence of the pH on adherence appears only in systems where dewetting occurs, above 45 phr loading. Clearly, the more cidic pigment surfaces have the least adherence to the elastomer with the vulcanisation system used, leading to more extensive vacuole ormation and increased swelling.

The reported dispersion studies indicate that under commonly used onditions, such as Banbury mixing, the dispersion of the silica is far rom perfect. The presence of oversize grains and of unwetted article clumps of the filler deteriorates the mechanical properties of he vulcanisates.

14.7 SUMMARY

The physical characteristics of silica pigments covering a wide range of properties were determined and related by regression analysis to he static and dynamic mechanical properties of SBR vulcanisates einforced with these silicas. Three morphological parameters were ound to determine the vulcanisates' characteristics: the external urface area by CTAB adsorption, the structural index, and the DBP absorption number.

The adherence between silica and elastomer was studied by means of swelling in solvents, by determination of the volume increase on leformation. It was found that dewetting, the formation of vacuoles on deformation, is a most important factor, limiting reinforcement by silica at loadings above 40 phr. Equally, at such loadings the pH of the silica has an important influence: the higher the acidity, the less the adherence between pigment particle and elastomer. Silane coupling agents not only increase adherence, but also improve the degree of dispersion of the silica, which is frequently characterised by the presence of large, partly unwetted particle clumps under conventional conditions of incorporation, thus severely limiting the strength properties of the vulcanisates. The curing system employed and the accelerators used are also of paramount significance for adherence

ACKNOWLEDGEMENT

The authors wish to thank the D.G.R.S.T. of the French Ministry of Industrial and Scientific Development for financial aid.

REFERENCES

1. J. B. DONNET and A. VOET. *Carbon Black*, 1976, Marcel Dekker, New York.
2. J. JANZEN and G. KRAUS. *Rubb. Chem. Tech.*, 1971, **44**, 1287.
3. B. C. LIPPENS and J. H. de BOER. *J. Catal.*, 1965, **4**, 319.
4. A. VOET and J. C. MORAWSKI. *Rubb. Chem. Tech.*, 1974, **47**, 758, 765.
5. A. VOET and W. N. WHITTEN, Jr. *Rubber World*, 1962, **146**(3), 77; 1963 **148**(5), 33.
6. G. KRAUS. *J. Appl. Poly. Sci.*, 1963, **7**, 861: *Rubb. Chem. Tech.*, 1964, 37 6.
7. V. DUCHACEK. *Angew. Makromol. Chem.*, 1972, **23**, 21.
8. J. P. FACHLER, J. A. FETCHIN and D. C. FRIES. *J. Am. Chem. Soc.* 1972, **94**, 7223.
9. J. R. WOLFE. *Rubb. Chem. Tech.*, 1968, **41**, 1339.
10. A. I. MEDALIA. *Rubb. Chem. Tech.*, 1974, **97**, 411.
11. E. PAPIRER, A. VOET and P. H. GIVEN. *Rubb. Chem. Tech.*, 1969, **42** 1200.

Chapter 15

EFFECT OF CARBON BLACK ON THE DYNAMIC PROPERTIES OF RUBBER

A. I. MEDALIA

15.1 INTRODUCTION

The effect of carbon black on the dynamic properties of rubber was a major interest of Dr Payne from the inception of his career. This subject is of vital importance in most applications of rubber, including tyres, engine mounts, bushings, and many other engineering applications. Payne's work dominated this field for over a decade, and in many aspects his results remain definitive [1, 2].

15.2 STATUS PRIOR TO PAYNE

Prior to Payne's researches, the principal phenomena associated with this topic had, in fact, been established, but mainly in a qualitative way or under restricted circumstances, and with inadequate or conflicting theoretical interpretations. These phenomena may be summarised as follows:

1. The elastic modulus of the filled compound (G_f') is greater than that of the gum (G_g') [3].
2. The loss tangent (tan δ) of the filled compound is greater than that of the gum [3–6].
3. G_f' depends on strain amplitude [3, 7, 8] ('amplitude effect').
4. Both the amplitude effect and the loss tangent increase with loading of carbon black [3, 4, 9–11], and reinforcing character (surface area) of carbon black [3, 6, 9–11].
5. The ratio of dynamic to static modulus is greater for filled compounds than for gum [12–14].

257

6. There is a general parallelism of electrical conductivity and dynamic modulus [15].

Another important effect of carbon black, broadening of the temperature or frequency dependence of dynamic properties, was described by Fletcher and Gent [16] just prior to Payne's own researches on this subject [17, 18].

Among the theories proposed to account for these phenomena, some were based on the hydrodynamic effect of the filler on the modulus [19, 20], some on colloidal effects (carbon black network formation) [11, 15, 21], and some on the altered behaviour of polymer molecules at or near the filler interface [11].

Many types of apparatus were in use for studying dynamic properties [2, 22, 23], including those based on free vibration, on resonant vibration, on coupled-mechanical oscillation, and on forced non-resonant oscillation, but none was completely satisfactory for the study of black-filled vulcanisates.

15.3   PAYNE'S RESEARCH CONTRIBUTIONS

15.3.1   Instrumental improvements

In the early 1950s Payne recognised the importance of the effect of carbon black on dynamic properties, and the need for an improved instrument for studying these properties. He realised that, in view of the effects of amplitude and frequency, instruments based on free or resonant vibration were of limited utility. He then showed [24] that the widely used Roelig machine [25] was subject to errors owing to coupling of the vibrations of the sample and the spring–mass system of the machine. The forced non-resonant type, which had been under development by Fletcher and Gent [26] and Davies [27], was the method of choice, but the instruments then in use relied on delicate transducer systems. Payne introduced the use of linear variable differential transformers (LVDTs) for both stress and strain measurement in a suitably rugged apparatus [28, 29] which, in various embodiments, is known as the RAPRA or Wallace–Payne machine. Special techniques for working at ultra-low amplitude, and for measuring phase angle, were introduced by Payne [28, 29]. This machine, with various modifications, was used not only in Payne's own classic researches, but also by other research workers in the 1960s and 1970s [30, 31].

15.3.2 Amplitude effect

Although Payne did not discover the effect of amplitude on the dynamic properties of carbon-black-filled vulcanisates, he perceived the fundamental importance of the amplitude effect and developed it as the basis for understanding many of the other effects of carbon black. Because of his detailed studies, it would be appropriate to refer to the amplitude effect (specifically, on the elastic modulus) as the Payne effect; however, the author's preference is for descriptive rather than personal nomenclature.

As shown [32] in Fig. 15.1, Payne's semi-logarithmic plots emphasised the apparent leveling-off of the elastic modulus at both low and high amplitude. (Others, however, have found a downturn in G' at extremely low amplitude [30, 33].) The difference between the

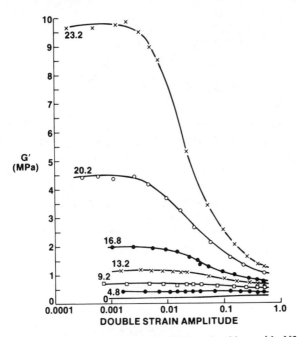

FIG. 15.1. Strain amplitude dependence of G' (butyl rubber with N330 black at indicated volume per cent) [32].

maximum or limiting value at low amplitude, G_0', and the limiting value at high amplitude, G_∞', can be referred to as $\Delta G'$, and is a quantitative measure of the amplitude effect.

In much of Payne's work he used very high loadings of carbon black in order to achieve a pronounced amplitude effect; but in the experiments of Fig. 15.1, a pronounced effect was achieved at normal loadings in butyl rubber, a polymer in which the carbon black–rubber interaction is low and the black–black interaction is correspondingly high [34]. Payne's interpretation of his results such as Fig. 15.1 is shown in Fig. 15.2, in which the author has replaced Payne's 'filler–

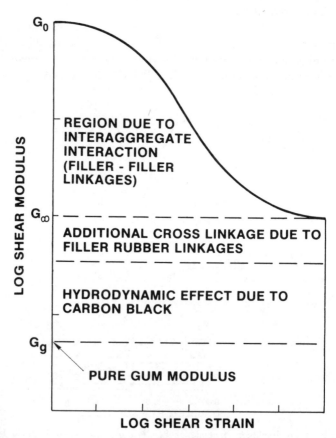

FIG. 15.2. Qualitative interpretation of strain amplitude dependence of G' (After Payne [32]).

ler linkages' by the term 'interaggregate interaction' in view of our
esent recognition of the aggregate nature of carbon black. In the
uthor's opinion [2], the contribution ascribed to filler–rubber linkages
really part of the hydrodynamic effect, since the hydrodynamic
eories are themselves based on bonding of the rubber to the filler.
Payne found that the dependence of elastic modulus on strain
mplitude (Fig. 15.3), or, as he preferred, on *strainwork* (Fig. 15.4),
uld be analysed quantitatively on the basis of a log-normal dis-
ibution of the strain amplitude, or strainwork, required to break the
iteraggregate bonds [32, 35, 36]. While he generally referred to this in
:rms of a log-normal distribution of bond strengths, he pointed out
at it could also signify that as large agglomerates break down to
naller ones, the smaller ones require more work for cleavage [36].
y normalising the elastic modulus as shown in Fig. 15.3 and 15.4,
ata at different loadings in different rubbers could all be treated in
e same manner. This quantitative treatment may be compared with
imilar qualitative concepts proposed earlier by Waring [15, 21] and
y Fletcher and Gent [37].
Payne's work provided the first clear-cut evidence that the hys-

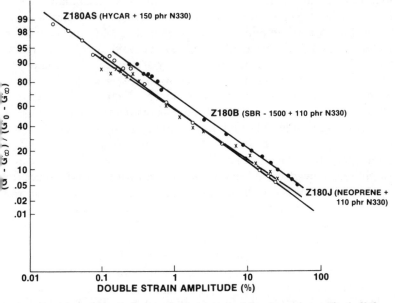

FIG. 15.3. Dependence of normalised elastic modulus on strain amplitude [36].

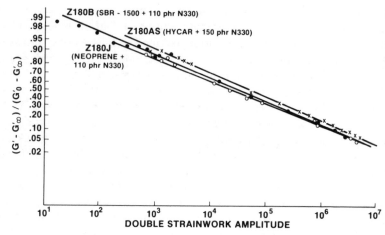

FIG. 15.4. Dependence of normalised elastic modulus on strainwork [36].

teretic parameters, G'' and δ (or tan δ), pass through a maximum wit
increasing amplitude (Fig. 15.5) [32]. He emphasised that the max
mum value of G'' occurs at the inflection point of G' on the sem
logarithmic plot. Note, however, that the loss compliance (J''),
hysteretic parameter of some practical interest, is a monotonicall
rising function of amplitude. He developed [36, 38] the relation be
tween G'' and G' in terms of the Cole–Cole [39] circular arc relatio
(Fig. 15.6) used for electrical permittivity and mechanical compliance
(Note the very high loadings of black used in this work.)

Qualitatively the amplitude dependence of tan δ was interprete
[40, 41] on the assumption that hysteresis results from breakdown an
re-formation of filler–filler linkages—processes of a retarding or hys
teretic nature. At low amplitudes there is little breakdown of th
interaggregate linkages, therefore little hysteresis. At intermediat
amplitudes considerable breakdown and re-formation take place, giv
ing high hysteresis. At high amplitudes the interaggregate networ
structure is broken down 'so extensively that re-formation of structur
is very much slower than the cycle time ' [40, 41]. This qualitativ
interpretation still appears to be essentially correct, but unfortunatel
has not yet been placed on a quantitative basis. Some experiment
questions also remain, including: (a) the observation [42] that in som
cases, tan δ (rather than G'') passes through its maximum at th
inflection point of G'; (b) the extent to which tan δ at very hig

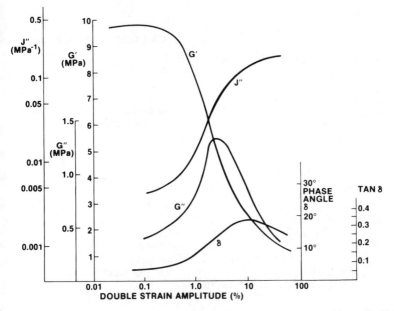

FIG. 15.5. Strain amplitude dependence of dynamic parameters (butyl rubber with 23·2 volume per cent of N330 black).

mplitudes approaches the value of the gum; and (c) the manner in which $\Delta G'$, and tan δ at high amplitudes, depend on frequency over a wide range.

5.3.3 Effect of carbon black

Payne's many contributions to our understanding of the effect of the nature and dispersion of carbon black on the dynamic properties of rubber stemmed principally from his understanding of the amplitude effect. He demonstrated [43] that short mixing times, which had been shown [44] to give a rough, mottled appearance of rubber sections examined by light microscopy, led to a large amplitude effect (large $\Delta G'$) and high tan δ. Further mixing, giving improved dispersion, reduced the amplitude effect and the hysteresis. This is of course attributable to reduction in the extent of interaggregate interaction as the agglomerates are broken down and the aggregates are more separated from each other.

In several papers Payne reported on the dynamic properties of systems in which the interaction between carbon black and rubber

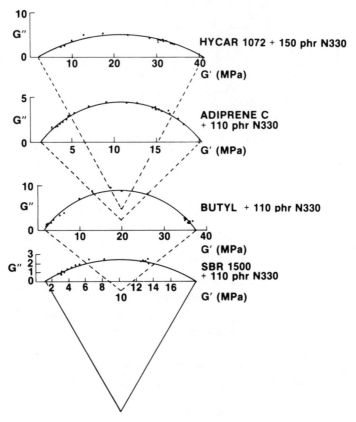

FIG. 15.6. Cole–Cole plots of G'' versus G' for various elastomer compounds [36].

was altered. Heat treatment of carbon black up to 2700°C ('graphi-tisation') was shown [45] to give a progressively larger amplitude effect and hysteresis, without significantly affecting G'_∞. Payne attri-buted this to poorer microdispersion of the black, owing to reduced rubber–black interaction during mixing. Subsequent work [46] has not shown any obvious difference in dispersion and has led to the suggestion that the interaggregate attractive forces are higher with heat-treated blacks.

Increased interaction between carbon black and butyl rubber [47, 48] or natural rubber (NR) [49] was achieved by several methods including heat treatment of the mix, use of chemical promoters, and

se of attrited carbon black. As expected, increased rubber–filler
nteraction gave a reduced amplitude effect and lower hysteresis.

Throughout his work Payne emphasised the general similarity of
the amplitude effect in various rubbers [36] and other systems [50],
although the *magnitude* of the effect depends on the rubber or other
vehicle and the state of dispersion. In a given rubber–black compound
the magnitude of the amplitude effect was essentially the same at four
different levels of cure [51]. However, there was no simple relation
for the combined effects of carbon black and crosslink density on G'_∞
and hysteresis.

Electrical conductivity in carbon-black-filled systems depends on
through-going paths of carbon black aggregates in close proximity.
These paths are disturbed by deformation, leading to cyclical variation in conductivity with the strain cycle, as shown by Waring [15]
and in more detail by Payne [52] and later by Voet and coworkers
[53, 54]. Payne's work on the temperature–frequency dependence of
dynamic properties [17, 18] confirmed and extended the work of
Fletcher and Gent [16], showing a generally satisfactory adherence to
the WLF equation with virtually no alteration of the reference temperature owing to the addition of carbon black, and showing
broadening of the dependence of dynamic properties on frequency or
temperature.

In a final contribution to this field, Payne and Whittaker [55]
showed that the hysteresis of carbon-black-filled vulcanisates followed a number of rules which had been established [56, 57] for
'domain' hysteresis in gas adsorption. The domains, which were
identified with a 'degenerated network' [55] or agglomerates, make a
transition in one direction (hard to soft) on the 'up' cycle of a
hysteresis loop, while the reverse transition is delayed until a lower
applied stress is reached on the 'down' cycle.

15.4 CURRENT STATUS

15.4.1 Carbon black properties affecting dynamic behaviour
Research in this field has continued since Payne's work and it seems
appropriate to summarise here our present knowledge and views. The
principal properties of carbon black are listed in Table 15.1. The
bulkiness of the individual aggregates, generally referred to as 'structure', and measured by vehicle absorption [58], is the property which

TABLE 15.1
Properties of carbon black affecting dynamic behaviour of rubber compounds

---

1. Aggregate bulkiness ('structure') (G'_∞)
2. Surface area ($\Delta G'$, tan δ)
3. Interaction with rubber ($\Delta G'$, tan δ)
    relative strength of rubber–filler versus filler–filler bonding
    dispersion
    breakdown of aggregates in rubber
    distribution in polymer blends
    cure
4. Aggregate morphology (fine tuning)
    aggregate size distribution ($\Delta G'$, tan δ)
    anisometry
    'openness,' amount of carbon per aggregate

---

almost uniquely determines G'_∞ at a given loading, as shown by the author [31, 46] in 1973. Surface area, on the other hand, plays the dominant rôle in the amplitude effect ($\Delta G'$) [31, 46] and hysteresis (expressed as tan δ) [59–61]. These effects are shown in Fig. 15.7 Payne [35] originally supposed that surface area controlled G'_0, but this is because the blacks available to him at the time were mostly of similar structure level, so that G'_0 was dominated by the amplitude effect.

The *magnitude* of the effects of 'structure' and surface area depends, of course, on loading, and also (especially the effect of surface area) on the interaction of the carbon black with rubber. The effects of relative bonding strength and of dispersion were elucidated by Payne, as described above. When carbon black is incorporated in rubber, some fracture of the aggregates can take place, to an extent which depends on both the black [62] and the rubber [63]. This fracture should affect the bulkiness of the aggregates and their contribution to G'_∞, as shown in detailed studies of rubber properties as a function of carbon black morphology measured in the vulcanisate [63]. The interaction of carbon black with different elastomers can affect its distribution in elastomer blends, which can apparently have a significant effect on the dynamic properties [64–66]. The effect of carbon black on state of cure is well known; tighter cure, of course gives higher elastic modulus and lower hysteresis (tan δ).

Returning to aggregate morphology, several more subtle aspects ('fine tuning') must be considered in addition to the major headings of 'structure' and surface area. Broader aggregate size distribution has

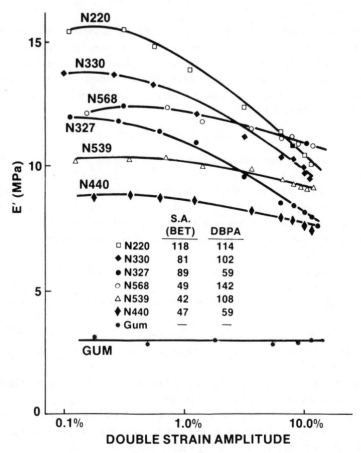

	S.A. (BET)	DBPA
□ N220	118	114
◆ N330	81	102
● N327	89	59
○ N568	49	142
△ N539	42	108
◆ N440	47	59
＊ Gum	—	—

FIG. 15.7. Strain amplitude dependence of E' (SBR-1500 with 50 phr of carbon black) [31].

een shown to give reduced tan δ [67–70], apparently owing to better acking of the aggregates. Anisometry of the aggregates has been laimed to have a significant effect on 'structure' in conjunction with ie effect of bulkiness [63, 71]. Finally, the 'openness' of the aggreates, which is related to the amount of carbon per aggregate at a iven level of 'structure' as measured by dibutylphthalate absorption DBPA) may have a slight effect: the more 'open' aggregates of the iew-technology' blacks tend to give a higher amplitude effect and lightly higher tan δ, owing to more extensive interaggregate contact 31, 34, 46, 72].

15.4.2 Theoretical

The principal line of theoretical development is based on a hydro
dynamic treatment of elastic modulus, and a colloidal treatment of th
amplitude effect and hysteresis. The hydrodynamic treatment is base
on Einstein's analysis of the effect of suspended spheres on th
viscosity of a liquid, adapted to higher concentrations as in th
Guth–Gold equation [73]

$$\eta = \eta_0(1 + 2\cdot5\phi + 14\cdot1\phi^2) \qquad (15.1)$$

(where η and η_0 are the viscosities of the filled and unfilled com
pound, respectively, and ϕ is the volume fraction of the filler) an
applied to modulus by substituting shear strain for shear rate. Wit
most carbon blacks the modulus (whether static or elastic) is highe
than predicted, to an extent which depends on the 'structure' of th
black. This is attributed [74] to occlusion of part of the rubber withi
the interstices of the aggregates. The occluded rubber is shielde
from deformation and thus acts as part of the filler. However, th
shielding is not fully effective and so an effectiveness factor [46, 7!
must be introduced, based on experimental measurement. If aggregat
bulkiness is measured by DBPA, and no allowance is made fo
breakdown on incorporation in rubber, an effectiveness factor o
0·5 gives good agreement [46] with experimentally measured G'_∞.

The colloidal treatment is based on interaggregate interaction and i
identified with the work of Payne, as discussed above. The amplitud
effect is attributed to breakdown of agglomerates, or a network, at th
extremes of strain cycling; and hysteresis is attributed to re-formatio
and re-breakdown of the agglomerates or network during cycling.

The hydrodynamic and colloidal treatments apply to differer
aspects of dynamic behaviour and may be joined together in a hydro
dynamic-occlusion-interaction (HOI) theory which provides
framework for treatment of all the effects of carbon black on th
dynamic properties of rubber [2]. Within this framework, reasonabl
qualitative explanations have been given for most observations, an
certain aspects have been analysed quantitatively, as discusse
above, or have been found to follow simple quantitative relation
[60, 61]. However, much remains to be done within the HOI theory
first, there exists considerable controversy regarding the basi
hydrodynamic equation for spheres at finite concentration and
would be desirable if this equation could be more firmly establishe
secondly, a theoretical calculation of shielding of stress by aggregate

hould be attempted; and thirdly, a quantitative theory of hysteresis
hould be developed.

Alternative theories have also been proposed. Aggregate
nisometry has been suggested as the basis for a modified hydro-
ynamic equation [20, 76], or for modifying the effectiveness of the
ccluded rubber [63, 71]. Rather than occlusion, a shell of im-
nobilised rubber has been suggested [77]; however, this would pre-
ict dependence of G'_∞ on surface area, rather than on structure as
bserved. A number of other alternative theories have been discussed
lsewhere [2].

From the experimental standpoint the trend is continuing toward
evelopment of improved instruments, giving greater range of
requency and amplitude, more precise measurement of dynamic
roperties, and more rapid and precise computation. In the 1940s and
950s, resonant instruments were favoured; in the 1960s, forced
on-resonant mechanical instruments were employed; and now,
ervo-controlled hydraulic or electromagnetic machines are available
ffering great flexibility and ease of operation. Digital read-out and
omputer calculation of the results will permit gathering of basic data
t a rate well in excess of our capacity to analyse and understand it.

New carbon blacks and other fillers continue to be developed. More
efined methods of characterising them may give better insight into
he relation between structure and properties. Study of different
lasses of polymers, curing systems, oil, and other additives may help
s understand the interaction of carbon black and vehicle.

From the theoretical standpoint much remains to be done, both
vithin the framework of the HOI theory, and perhaps along other
heoretical lines. From a practical standpoint we may look for more
letailed computer analysis of the stress–strain behaviour of rubber
rticles under dynamic conditions, leading to a more sophisticated
pplication of basic dynamic data to engineering applications (another
ubject of great interest to Dr. Payne), and a more realistic assess-
nent of the dynamic requirements of the filled vulcanisate.

REFERENCES

1. A. R. PAYNE and R. E. WHITTAKER. *Rubb. Chem. Tech.*, 1971, **44**, 440.
2. A. I. MEDALIA. *Rubb. Chem. Tech.*, 1978, **51**, 437.
3. S. D. GEHMAN, D. E. WOODFORD and R. B. STAMBAUGH. *Ind. Eng. Chem.*, 1941, **33**, 1032.

4. J. H. FIELDING. *Ind. Eng. Chem.*, 1937, **29**, 880.
5. H. ROELIG. (1938). In: *Proceedings Rubber Technology Conference*, London, p. 821.
6. D. PARKINSON. *Trans. Inst. Rubb. Ind.*, 1940, **16**, 87.
7. W. J. S. NAUNTON and J. R. S. WARING. *Trans. Inst. Rubb. Ind.*, 193?, **14**, 340.
8. R. B. STAMBAUGH. *Ind. Eng. Chem.*, 1942, **34**, 1358.
9. D. PARKINSON. *Trans. Inst. Rubb. Ind.*, 1943, **19**, 131.
10. D. PARKINSON. *Trans. Inst. Rubb. Ind.*, 1945, **21**, 7.
11. K. E. GUI, C. S. WILKINSON, JR and S. D. GEHMAN. *Ind. Eng. Chem* 1952, **44**, 720.
12. J. H. DILLON, J. B. PRETTYMAN and G. L. HALL. *J. Appl. Phys.*, 194?, **15**, 309.
13. C. W. KOSTEN. (1938). In: *Proceedings Rubber Technology Conference*, London, p. 987.
14. L. MULLINS. *Trans. Inst. Rubb. Ind.*, 1950, **26**, 27.
15. J. R. S. WARING. *Ind. Eng. Chem.*, 1951, **43**, 352.
16. W. P. FLETCHER and A. N. GENT. *Brit. J. Appl. Phys.*, 1957, **8**, 194.
17. A. R. PAYNE. (1958). In: *The Rheology of Elastomers* (Eds P. MASON and N. WOOKEY), p. 86. London: Pergamon.
18. A. R. PAYNE. (1959). In: *The Physical Properties of Polymers*, p. 27? London: Society of Chemical Industry. New York: Macmillan.
19. H. M. SMALLWOOD. *J. Appl. Phys.*, 1944, **15**, 758.
20. E. GUTH. *J. Appl. Phys.*, 1945, **16**, 20.
21. J. R. S. WARING. *Trans. Inst. Rubb. Ind.*, 1950, **26**, 4.
22. S. D. GEHMAN. *Rubb. Chem. Tech.*, 1957, **30**, 1202.
23. A. C. EDWARDS and G. N. S. FARRAND. (1961). In: *The Applied Science of Rubber* (Ed. W. J. S. NAUNTON), Ch. 8. London: Edward Arnold.
24. A. R. PAYNE. (1956). In: *Proceedings 3rd Rubber Technology Conference*, London, pp. 413.
25. H. ROELIG. (1938). In: *Proceedings Rubber Technology Conference*, London, p. 821. H. ROELIG. *Kautsch.*, 1943, **19**, 47.
26. W. P. FLETCHER and A. N. GENT. *J. Sci. Instrum.*, 1952, **29**, 186.
27. D. M. DAVIES. *Engineering*, 1953. **176**, 196.
28. A. R. PAYNE. *Rev. Gen. Caout.*, 1956, **33**, 913.
29. A. R. PAYNE. *Mat. Res. Stand.*, 1961, **1**, 942.
30. A. VOET and F. R. COOK. *Rubb. Chem. Tech.*, 1967, **40**, 1364.
31. A. I. MEDALIA. *Rubber World*, 1973, **168**(5), 49.
32. A. R. PAYNE. *Rubb. Plast. Age*, 1961 (August), 963.
33. A. VOET and F. R. COOK. *Rubb. Chem. Tech.*, 1968, **41**, 1215.
34. A. I. MEDALIA and S. G. LAUBE. *Rubb. Chem. Tech.*, 1978, **51**, 89.
35. A. R. PAYNE. *J. Appl. Poly. Sci.*, 1962, **6**, 57.
36. A. R. PAYNE. *J. Appl. Poly. Sci.*, 1964, **8**, 2661.
37. W. P. FLETCHER and A. N. GENT. *Trans. Inst. Rubb. Ind.*, 1953, **29**, 266
38. A. B. DAVEY and A. R. PAYNE. (1964). *Rubber in Engineering Practice*, London: Maclaren. New York: Palmerton.
39. K. S. COLE and R. H. COLE. *J. Chem. Phys.*, 1941, **9**, 341.
40. A. R. PAYNE. *Rubb. J.*, 1964, **146**, 36.

1. A. R. PAYNE. (1965). In: *Reinforcement of Elastomers* (Ed. G. KRAUS), Ch. 3. New York: Interscience.
2. L. R. BARKER, A. R. PAYNE and J. F. SMITH. *J. Inst. Rubb. Ind.*, 1967, **1**, 206.
3. A. R. PAYNE. *J. Appl. Poly. Sci.*, 1965, **9**, 2273.
4. B. B. BOONSTRA and A. I. MEDALIA. *Rubb. Age*, 1963, **46**, 892.
5. A. R. PAYNE. *J. Appl. Poly. Sci.*, 1965, **9**, 3245.
6. A. I. MEDALIA. *Rubb. Chem. Tech.*, 1973, **46**, 877.
7. A. R. PAYNE. *J. Appl. Poly. Sci.*, 1963, **7**, 873.
8. A. M. GESSLER and A. R. PAYNE. *J. Appl. Poly. Sci.*, 1963, **7**, 1815.
9. A. R. PAYNE, P. M. SWIFT and M. A. WHEELANS. *J. Rubb. Res. Inst. Malaya*, 1969, **22**, 275.
10. A. R. PAYNE. *J. Coll. Sci.*, 1964, **19**, 744.
11. A. R. PAYNE, R. E. WHITTAKER and J. F. SMITH. *J. Appl. Poly. Sci.*, 1972, **16**, 1191.
12. A. R. PAYNE. *J. Appl. Poly. Sci.*, 1965, **9**, 1073.
13. A. VOET and F. R. COOK. *Rubb. Chem. Tech.*, 1968, **41**, 1207.
14. A. VOET and J. C. MORAWSKI. *Rubb. Chem. Tech.*, 1974, **47**, 765.
15. A. R. PAYNE and R. E. WHITTAKER. *Trans. Faraday Soc.*, 1970, **66**, 2383.
16. D. H. EVERETT and W. I. WHITTON. *Trans. Faraday Soc.*, 1952, **48**, 749.
17. J. A. ENDERBY. *Trans. Faraday Soc.*, 1955, **51**, 835.
18. ASTM Test D2414-76.
19. D. PARKINSON. *Trans. Inst. Rubb. Ind.*, 1940, **16**, 87.
20. G. KRAUS and J. JANZEN. *Kaut. und Gummi Kunst.*, 1975, **28**, 253.
21. J. M. CARUTHERS, R. E. COHEN and A. I. MEDALIA. *Rubb. Chem. Tech.*, 1976, **49**, 1076.
22. F. A. HECKMAN and A. I. MEDALIA. *J. Inst. Rubb. Ind.*, 1969, **3**, 66.
23. G. C. McDONALD and W. M. HESS. *Rubb. Chem. Tech.*, 1977, **50**, 842.
24. W. M. HESS, C. E. SCOTT and J. E. CALLAN. *Rubb. Chem. Tech.*, 1967, **40**, 371.
25. A. K. SIRCAR and T. G. LAMOND. *Rubb. Chem. Tech.*, 1973, **46**, 178.
26. A. K. SIRCAR, T. G. LAMOND and P. E. PINTER. *Rubb. Chem. Tech.*, 1974, **47**, 48.
27. J. JANZEN and G. KRAUS. (1972). In: *Proceedings International Rubber Conference*, Brighton, UK, 1972, p. G7-1.
28. C. J. STACY, P. H. JOHNSON and G. KRAUS. *Rubb. Chem. Tech.* 1975, **48**, 538.
29. W. M. HESS and V. E. CHIRICO, *Rubb. Chem. Tech.*, 1977, **50**, 301.
30. G. KRAUS. (1977). In: *Proceedings International Rubber Conference*, Brighton, UK, 1977, paper 21.
31. J. D. ULMER, W. M. HESS and V. E. CHIRICO. *Rubb. Chem. Tech.*, 1974, **47**, 729.
32. A. I. MEDALIA, E. M. DANNENBERG, F. A. HECKMAN and G. R. COTTEN. *Rubb. Chem. Tech.*, 1973, **46**, 1239.
33. E. GUTH and O. GOLD. *Phys. Rev.*, 1938, **53**, 322. E. GUTH. (1938). In: *Proceedings 5th International Congress Applied Mechanics*, Cambridge, 1938, p. 448.

74. A. I. MEDALIA. *J. Coll. Interf. Sci.*, 1970, **32**, 115.
75. A. I. MEDALIA. *Rubb. Chem. Tech.*, 1972, **45**, 1171.
76. G. B. JEFFREY. *Proc. Royal Soc. (London)*, 1922, **A102**, 161.
77. P. P. A. SMIT. *Rheol. Acta*, 1966, **5**, 277. P. P. A. SMIT. *Rheol. Act* 1968, **8**, 277.

Chapter 16

EFFECT OF CROSSLINK TYPE ON THE PERFORMANCE OF RUBBER AS AN ENGINEERING MATERIAL

E. SOUTHERN

16.1 INTRODUCTION

he discovery by Goodyear [1] and Hancock [2] of a method of osslinking natural rubber transformed it from an interesting cientific curiosity into an engineering material, as the wide range of pplications described in their works clearly demonstrates. It is erhaps a little surprising that, some 150 years later, the most popular ethod of crosslinking using sulphur is still essentially the same. The se of accelerators has reduced both the amount of sulphur required d, more importantly, the vulcanisation time considerably as shown Table 16.1. More recent developments have caused an increase in corch and cure time at 140°C but these systems are designed for use much higher temperatures (150–180°C) where adequate scorch fety is very important because the vulcanisation time is so short.

The sulphur crosslinking systems produce crosslinks containing etween one and five or six sulphur atoms. During the vulcanisation rocess, some of the sulphur is expelled from the initial polysulphidic osslinks which may be reduced eventually to monosulphidic cross-ks. In a conventional sulphur vulcanisate of natural rubber con-ining 2·5 parts by weight of sulphur and 0·5 parts of accelerator per undred of rubber, the relative proportions (per cent) of crosslink pe are: polysulphidic, 70; disulphidic, 15; and monosulphidic, 15. By creasing the accelerator-to-sulphur ratio and retaining a relatively ng cure time a vulcanisate is obtained which contains predominantly nore than 90 per cent) monosulphidic crosslinks. Such a vulcanisate called an efficiently vulcanised (EV) system. In this context,

273

TABLE 16.1
Development of vulcanisation natural rubber [3]

Date	Formulation		Cure time at 140°C
1840	rubber	100	6 hr
	sulphur	8	
1850	rubber	100	
	sulphur	8	4 hr
	metal oxide	5	
1906	rubber	100	
	sulphur	6	2 hr
	ZnO	5	
	Acc. TC	2	
1921	rubber	100	
	ZnO	5	
	S	3	20 min
	MBT	1	
	stearic acid	1	
1920s	early heat-resistant systems (very scorchy, *e.g.*, TMT)		15 min
1940s	delayed action systems		30 min
1950s	EV systems (low sulphur/ high accelerator ratio)[a]		60 min
1969	soluble EV systems[a]		60 min
1970	soluble zinc soaps		

[a] EV, efficiently vulcanised.

'efficiently' refers to the use of sulphur, not to the cost. A convenient guide [4] to the relative amounts of sulphur and accelerator which must be used to produce EV and semi-EV vulcanisates is given in Table 16.2. It must be emphasised that the correct ratio alone is not

TABLE 16.2
Classification of sulphur vulcanisates [4]

System	Sulphur (pphr)	Accelerator (pphr)
Conventional	2·0–3·5	1·0–0·4
Semi-efficiently vulcanised	1·0–1·7	2·5–1·0
Efficiently vulcanised (EV)	0·3–0·8	6·0–2·5

Compound formulations and physical properties of urethane crosslinked gum natural rubber

Base mix: SMR 5L, 100; ZMBI, 2; ZDMC, 2; Caloxol W5G, 3; Flectol H, 2.

Compound No.	1	2	3	4	5	6	7	8	9
Novor 924	4·5	6·7	9·3	10·1	12·0	13·4	7	8	9
Novor 220							5	7	9
Cure	60 min at 160°C						30 min at 150°C		
Properties									
Hardness (IRHD)	<28	31	37	39·5	42	42·5	29	34·5	38
MR 100 (MPa)	–	0·465	0·715	0·711	0·794	0·855	–	0·556	0·680
$M_c^{-1} \times 10^5$	3·13	5·67	10·0	11·0	13·5	14·0	4·02	5·45	8·13
TS (MPa)	8·7	14·6	18·5	16·8	17·2	17·3	14·4	20·0	18·3
EB (%)	734	716	650	585	558	538	790	756	644
Resilience (%)[a]	67	74·5	82	82	84	84	69	76	78
SR rate (% per decade)	–	4·2	2·7	2·5	2·7	2·4	–	5·3	4·0
High strain set (%)	–	12	7	9	11	–	7	7	7
Aged 70 hr at 100°C (% retained)									
TS	–	–	93	–	58	–	–	93	–
EB	–	–	96	–	83	–	–	86	–
Aged 7 days at 100°C (% retained)									
TS	75	48	40	39	37	26	74	54	67
EB	94	85	79	84	81	64	84	76	84
Aged 14 days at 100°C (% retained)									
TS	26	12	10	11	11	11	47	24	18
EB	67	53	47	47	46	41	70	54	43

MR 100, relaxed modulus at 100% strain; M_c, molecular weight between crosslinks; TS, tensile strength; EB, elongation at break; SR, stress relaxation.
[a] Dunlop tripsometer.

TABLE 16.4
Compound formulations and physical properties [41]

Compound No.	10 Conventional sulphur	11 Semi-EV	12 EV	13 EV	14 Urethane	15 Urethane
SMR 5	100	100	100	100	100	100
SMR 5 CV	50	50	50	50	50	50
HAF black	4	4	4	4		
Process oil[a]						
Peptiser[b]						
Zinc oxide	5	3·5	5	5	2	2
Stearic acid	3	2·5	2	2	5	5
Drying agent[c]					1	1
ZDMC					2	3
TBBS	0·5					2
Sulphur	2·5	1·2	0·33	0·4		
Novor 924 (urethane)[d]					6·7	6·7
Antioxidant[e]	2	2	2	2	2	2
CBS		0·8	2			
TMTD		0·4	3			
MBS[f]				6		
Mooney scorch, t_s at 120°C, min	24	13	9·5	25	13	12
Cure time/temp., min/°C	40/140	30/140	40/140	60/140	27/153	27/153

Initial vulcanisate properties					
Hardness, IRHD	65	65	67	60	62	67·5
MR 100, MPa[g]	2·08	2·22	2·34	1·81	1·69	2·34
M 300, MPa	14·3	18·2	17·3	13·3	12·7	15·8
TS, MPa	28·6	30·1	24·2	24·6	19·2	20·4
EB, %	513	483	389	444	430	384
Dunlop resilience, %	70	77	67	63	56	60
Ring fatigue at 100% extn., kc	223	106	68	64	160	94
Goodrich heat build-up, °C	29	32	36	40	89	81
Permanent set, %	3·2	1·2	1·0	1·3	24	32
Compression set (25%, 24 hr at 70°C),%	27	14	10	13	34	31
Aged vulcanisate properties					
Retention of TS after 7 days at 100°C, %	27	46	76	87	52	
Retention of TS after 70 hr at 125°C, %	21	18	50	38	6	12
Revision after 1 hr at 180°C on rheometer, %	48	39	5	15	2·5	

[a] Dutrex 729 (Shell).
[b] Ancoplas ER (Anchor Chemical Co.).
[c] Caloxol W5G (John Sturge).
[d] Durham Chemicals.
[e] Santoflex 13 (Monsanto).
[f] Morpholino-2-benzothiazole sulphenamide.
[g] 1 MPa = 10·2 kgf/cm² = 145 p.s.i.

TABLE 16.5
Compound formulations and physical properties [22]

Formulation component/property	Compound No.					
	16	17	18	19 Soluble EV	20 EV	21 Peroxide
	Conventional					
	A	B	C	D	E	F
Component						
SMR 5	100	100	100	100	100	100
Zinc oxide	5	5	5	5	5	5
SRF black (N 762)	50					
FEF black (N 550)		45	45	50	50	50
Naphthenic oil	10	10	10			
Stearic acid	2	2	2			
Zinc 2-ethylhexanoate				1	2	
Antioxidant	2[a]	1·5[b]	1·5[b]	2[b]	2[b]	2[b]
Antiozonant[c]		1·5	1·5			
Sulphur	2·5	2·25	2·25	0·6	0·25	
DTM[d]		1·0	1·0			
MBTS	1·0	1·0	1·0			
DPG	0·1	0·2	0·2			
OBS				1·5	2·1	
TBTD				0·6		
TMTD					1·0	

Hardness, IRHD	54	58	64	59	61	57
TS, MPa	24	25	26	25	25	22
EB, %	570	520	475	500	505	330
Aged[e] 5 days at 100°C						
Change in hardness, IRHD	+6	+5	+6	+5	+3	−3
Change in TS, %	−65	−41	−46	−14	−15	−42
Change in EB, %	−57	−44	−53	−17	−13	−22
Aged[e] 5 days at 125°C						
Change in hardness, IRHD	+3	+5	+2	−3	−3	−9
Change in TS, %	B	B	B	−80	−69	−52
Change in EB, %	B	B	B	−62	−38	−15
Compression set[f], %						
22 hr at 70°C	29	26	21	13	10	5
22 hr at 100°C	48	46	51	31	18	9
22 hr at 125°C	60	63	63	42	26	15
Stiffness and dynamic properties						
MR 100, MPa	1·4	1·7	2·4	1·6	1·7	2·0
Resilience[g] at 50°C, %	87	85	88	85	78	86
Goodrich flexometer, heat build-up, °C	F	F	35	25	27	27
Goodrich flexometer, set, %			28	12	8	8

[a] Blend of aryl amines.
[b] TMQ.
[c] HPPD.
[d] N,N'-dithiobismorpholine.
[e] Aged in air oven.
[f] BS 903:A6.
[g] Dunlop tripsometer.
F, Failed by blow-out.
B, Brittle.

TABLE 16.6

Compound formulations and crosslink type [14]

	Compound No.						
	22 A	23 B1	24 B2	25 C	26 D	27 E	28 F
RSS 1	100	100	100	100	100	100	100
Zinc oxide	3.0	5.0	5.0	5.0	5.0	3.0	5.0
Stearic acid	–	1.5	1.0	–	1.5	–	1.5
Antioxidant[a]	1.5	1.5	1.5	1.5	1.5	1.5	1.5
Dicumyl peroxide	2.0	–	–	–	–	–	–
TMTD[b]	–	4.0	6.3	–	–	–	–
CBS[c]	–	–	–	1.0	0.5	–	0.54
ZDMC[d]	–	–	–	–	–	3.0	–
Aldehyde amine accelerator[e]	–	–	–	–	–	15.0	–
Sulphur	–	–	0.44	4.0	2.5	–	2.7
Cure conditions, time (min)	50[f]	120	120	120	1000	60	35
temperature (°C)	150	140	140	140	140	100	140
$(2 M_c)^{-1} \cdot 10^5$	4.62	5.30	4.70	4.80	3.40	1.10	5.20

Structural Analysis:

% Total crosslinks S_x ($x > 2$)	0	0	0	0	0	>90	ca. 70
% Total crosslinks S_2	0	<10	<10	<10	<10	<10	ca. 15
% Total crosslinks S_1	0	>90	>90	>90	>90	0	ca. 15
% Total crosslinks C–C	100	0	0	0	0	0	0
E	–	3·5	3·9	21	16	8·5	13
E'	–	3·5	3·7	20	15	1·5	7
Predominant main chain[g]	–	P	P	CS	CS	CS	CS
Modification type							

[a] Flectol H (Monsanto Chemicals Ltd) used.
[b] TMTD, Tetramethylthiuram disulphide.
[c] CBS, N-Cyclohexyl-2-benzthiazole-2-sulphenamide.
[d] ZDMC, Zinc dimethyldithiocarbamate.
[e] Vulcafor EFA (ICI Ltd) used.
[f] Precure of 10 min/110°C used to avoid anisotropy.
[g] CS = Cyclic sulphide along main chain.
P = Pendent accelerator fragment.

TABLE 16.7
Compound formulations and physical properties [41]

Base mix: RSSl 100; zinc oxide 4; Flectol H 2

	Compound No.				
	29	30	31	32	33
	Conventional high sulphur system	EV systems			
		1	2	3	4
Ingredients					
Lauric acid[a]	1·0	3·0	3·0	3·0	3·0
Sulphur	2·5	0·53	0·53	–	–
Sulfasan R	–	–	–	1·97	1·97
TMTD	–	1·0	1·2	1·2	–
MOR	–	2·1	–	–	–
MBTS	–	–	1·1	1·1	2·77
CBS	0·5	–	–	–	–
Mooney scorch 120°C, t_s, min	45	36	20	16	30
Cure time at 140°C, min					

Vulcanisate properties					
IRHD	39	44	43	48	44
Initial tensile strength, kg·cm^{-2}	244	239	237	226	233
Per cent retained after air oven ageing:					
7 days at 100°C	6	77	48	67	69
14 days at 100°C	untestable	68	39	45	61
Initial elongation at break, %	730	633	635	555	645
Per cent retained after air oven ageing:					
7 days at 100°C	23	92	76	93	89
14 days at 100°C	untestable	86	75	88	82
Initial M 300, kg·cm^{-2}	21	22	22	29	21
Per cent retained after air oven ageing:					
7 days at 100°C	(150 after 3 days)	109	136	110	114
14 days at 100°C	untestable	104	123	103	124
Compression set, % 24 hr at 70°C under 25 per cent strain	28	19	18	18	13

[a] This can be replaced by its equivalent of stearic acid.

TABLE 16.8
Compound formulations and physical properties [41]

Ingredients	Conventional high sulphur system	EV systems 1	EV systems 2	EV systems 3	EV systems 4	HAF black conventional	HAF black EV	FEF black conventional	FEF black EV	Lamp black conventional	Lamp black EV
RSS 1	100	100	100	100	100	100	100	100	100	100	100
Filler	–	–	–	–	–	50	50	50	50	50	50
Dutrex R	–	–	–	–	–						
Zinc oxide	4·0	4·0	4·0	4·0	4·0	5·0	5·0	5·0	5·0	5·0	5·0
Lauric acid[a]	1·0	3·0	3·0	3·0	3·0	5·0	5·0	5·0	5·0	5·0	5·0
Flectol H	2·0	2·0	2·0	2·0	2·0	1·0	3·0	1·0	3·0	1·0	3·0
Sulphur	2·5	0·53	0·53	–	–	2·0	2·0	2·0	2·0	2·0	2·0
Sulfasan R	–	–	–	1·97	1·97	2·5	0·25	2·5	0·25	2·5	0·25
TMTD	–	1·0	1·2	1·2	–	–	–	–	–	–	–
MOR	–	2·1	–	–	–	–	1·2	–	1·2	–	1·2
MBTS	–	–	1·1	1·1	2·77	–	–	–	–	–	–
CBS	0·5	–	–	–	–	0·4	1·8	0·4	1·8	0·4	1·8
Mooney scorch 120°C, t_s, min	45	36	20	16	30	21	17·5	28	19·5	32	22
Cure time at 140°C, min	40	40	40	40	60	40					

Vulcanisate properties											
IRHD	39	44	43	48	44	67	67	66	68	65	62
Initial tensile strength, kg·cm⁻²	244	239	237	226	233	271	260	252	230	210	207
Per cent retained after air oven ageing:											
7 days at 100°C	6	77	48	67	69	23	80	24	79	31	74
14 days at 100°C	untestable	68	39	45	61	15	68	18	65	21	67
Initial elongation at break, %	730	633	635	555	645	490	495	495	470	485	480
Per cent retained after air oven ageing:											
7 days at 100°C	23	92	76	93	89	22	71	25	84	33	87
14 days at 100°C	untestable	86	75	88	82	9	68	14	81	14	85
Initial M 300, kg·cm⁻²	21	22	22	29	21	149	134	142	128	120	110
Per cent retained after air oven ageing:											
7 days at 100°C	(150 after 3 days)	109	136	110	114	–	129	–	102	–	92
14 days at 100°C	untestable	104	123	103	124	–	112	–	87	–	87
Compression set, % 24 hr at 70°C under 25 per cent strain	28	19	18	18	13	31	17	30	13	21	13

aThis can be replaced by its equivalent of stearic acid.

sufficient to produce an EV vulcanisate with monosulphidic cross-links. A careful choice of accelerator and activators is necessary in order that the intermediate compounds formed are soluble at vul-canising temperatures and it is also necessary that the cure time is relatively long so that all the initially formed polysulphidic linkages are converted to monosulphidic crosslinks. Besides producing cross-links, the sulphur may also introduce main-chain modifications either in the form of pendent groups or as cyclic groups. These groups may influence the behaviour of the vulcanisate and techniques have been developed to estimate the amount of sulphur used in this way [5–9].

The other main crosslink type is the direct carbon–carbon crosslink which is usually achieved nowadays by the use of dicumyl peroxide [9]. Other peroxides [10] have been used and also high energy radiation [11], but these are more of scientific rather than commercial interest. Recently [12] a new crosslinking system, which has excep-tionally good reversion resistance, has been developed using urethane crosslinks. Some new data are presented in Table 16.3 for gum compounds using this system for comparison with the other cross-linking systems.

The compounds used for engineering applications are usually filled with carbon black. The amount of filler is normally relatively low and the particle size is in the medium range (*e.g.* FEF and SRF blacks). In some instances data for blacks other than these are quoted because of lack of information in the literature.

The most important properties for engineering applications can be conveniently divided into four main categories: strength, relaxation processes, energy loss, and ageing behaviour. While these groups do not contain all the properties which may be important for a particular application it is felt that they are sufficiently comprehensive for the framework of this paper. Full details of all the compounds and their physical properties are given in the Tables 16.4–16.8.

16.2   STRENGTH

16.2.1   Tensile strength

The tensile strength of rubbers is probably measured more frequently than any other property except hardness and yet rubbers are hardly ever used in applications where tensile strength is the limiting factor. Rubber bands are an obvious exception. The reason for its inclusion in specifications is that it is a good measure of compound quality and monitors the introduction of low-cost fillers. Engineering compounds are generally of high quality, so that the tensile strength forms an important part of the specification of an engineering compound.

The tensile strength of rubbers is also of considerable scientific interest and both crosslink type and degree of crosslinking have a significant effect on the tensile strength of natural rubber [13, 14] as shown in Fig. 16.1. In all cases the tensile strength passes through a maximum with increasing degree of crosslinking. The polysulphide crosslinked rubber (conventional vulcanisate) has the highest strength followed by monosulphide and urethane crosslinked material and the carbon–carbon crosslinked rubbers are the poorest. The curves for different types of crosslink show that the higher maximum values of the tensile strength occur at higher crosslink densities. Experimental

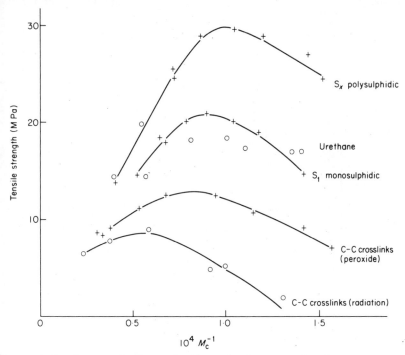

FIG. 16.1. Tensile strength of natural rubber vulcanisates as a function of crosslink density. M_c = molecular weight between crosslinks. (Data from ref. 13 except for urethane.)

evidence has been obtained [15] showing that the maximum in tensile strength occurs at approximately the same degree of crosslinking as the amount of crystallisation estimated from density measurements during extension. It has been suggested [16, 17] that the tensile strength decreases at high degrees of crosslinking owing to the breaking point being reached before the extension is high enough for crystallisation to occur in the bulk of the rubber. At low degrees of crosslinking it has been proposed [16] that plastic flow of the chains prevents crystallites from forming. A quantitative relationship has been derived [17] on the assumption that the tensile strength is governed by the fraction of the material which is capable of orientation and hence crystallisation.

More comprehensive data [14] on the effect of curing systems show that the shape of the curves is influenced not only by the crosslink type but also by the main-chain modifications as shown in Fig. 16.2. It is suggested that the vulcanisates with highly modified chains do not

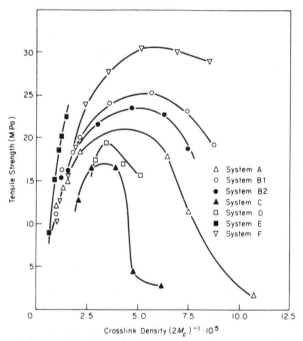

FIG. 16.2. Tensile strength of natural rubber vulcanisates as a function of crosslink density [14]. See Table 16.6 for explanation of key.

:rystallise as readily as those with less modified chains and con-
equently have lower strength. However, the vulcanisates with the
east modified chains which also have monosulphidic crosslinks have
ower tensile strengths than the conventional system which has pre-
lominantly polysulphidic crosslinks. Direct evidence of the greater
trength of vulcanisates with polysulphidic crosslinks was obtained
y treating a conventional vulcanisate with triphenylphosphine (TPP)
vhich desulphurates all the crosslinks to produce monosulphidic
inkages. As a result of this treatment the tensile strength of the
:onventional vulcanisate fell from 306 kg·cm^{-2} to 191 kg·cm^{-2}, thereby
lemonstrating in a convincing manner the superiority of the polysul-
hidic crosslinks for high tensile strength. Similar measurements
:arried out independently [18] have not confirmed these results.

The tensile strengths of vulcanisates are ranked in inverse order to
he bond energies of the crosslinks [13] and the suggested mechanism
or low-strength crosslinks giving high-strength rubbers is that the
veak crosslinks are mechanically labile, thereby acting as stress
elievers [19]. When a stress is applied to a rubber sample, the shorter
:hains will reach breaking point before the longer chains and it is
uggested that the polysulphide crosslinks break, thereby allowing the
:hains to extend further without breakage. The broken crosslinks will
e able to reform if they come into contact with suitable groups as the
:xtension proceeds. In effect, the load is more evenly distributed over
larger number of network chains. None of the other crosslinks,
nonosulphidic, urethane, or carbon–carbon, are capable of reforming
ince broken so, although they are stronger than polysulphidic cross-
inks, the stress-relieving mechanism is absent and, when a chain
reaks, the load is transmitted to neighbouring chains, some of which
ilso break. The process continues in this manner until the sample
reaks.

This view has been challenged [20] on the grounds that there is no
:vidence from stress relaxation data, carried out at various extensions
ip to break, that polysulphide crosslinks are mechanically labile at
oom temperature. It was suggested that the high strength of vul-
:anisates containing polysulphidic crosslinks was due to the labile
nature of these crosslinks at vulcanising temperatures, giving rise to a
more 'relaxed' network. A somewhat different view, but again oppos-
ing the stress relief mechanism, proposes [21] that whilst there are
good reasons for supposing that carbon–carbon crosslinks are randomly distributed on the main chains, the mechanism of sulphur

crosslinking will ensure that sulphur crosslinks are not randomly distributed. It is suggested that clustering will occur and that this behaviour will lead to higher strength. Both these arguments rely on the experimental evidence [18] cited earlier which indicates that there is no change in tensile strength when polysulphidic crosslinks are converted chemically to monosulphidic crosslinks. Since these data are in conflict with other published work [14] it would seem that further work is necessary to produce experimental evidence which is not disputed. A different experimental approach using EPDM rubber has been used to investigate the relative influence of crosslink distribution and crosslink type [11]. It was suggested that the very reactive termonomer (ethylidene norbornene) used in this rubber ensures that crosslinks are placed at the sites of this monomer and are not determined by the crosslinking mechanism. There was little effect of crosslink type but, unfortunately, as carbon-black-filled compounds were used in these studies and this filler is known to reduce the influence of crosslink type, the results are not as convincing as they would have been in a gum rubber.

As indicated, the effect of crosslink type in carbon-black-filled vulcanisates is much reduced. A conventional vulcanisate of natural rubber has a lower strength than the corresponding gum compound whereas the carbon black increases the strength of the EV compound so that they are not significantly different, as shown in Table 16.9. The filler also improves the strength of the peroxide-cured vulcanisate considerably so that it is almost as strong as the sulphur crosslinked compounds [22]. Relatively small differences are also observed in filled synthetic rubbers between sulphur crosslinks and carbon–

TABLE 16.9
Tensile strength of black filled natural rubber [22, 42]^a

Compound No.	16	10	19	12	15	21
Vulcanisation	C^b	C^b	EV^c	EV^c	urethane system	peroxide
Black type	SRF	HAF	SRF	HAF	HAF	SRF
Tensile strength, TS (MPa)	24	29	25^d	24	20	22

^{a} See Table 16.5 for formulation details.
^{b} C = conventional.
^{c} Efficiently vulcanised.
^{d} Soluble compounding ingredients.

arbon crosslinks but unfortunately the corresponding values of the ensile strength for unfilled rubbers were not measured [12].

6.2.2 Tear resistance

he tensile strength, tear, crack growth, and fatigue properties of ubbers are all manifestations of the general strength behaviour of ubbers. From a practical point of view the tear resistance is probably nly important with respect to accidental damage to the component. onventionally measured tear strengths (*e.g.*, crescent tear) are not ery useful since they are not simply related to the fundamental trength properties of the rubber. Although it is well known that the ear properties are strongly influenced by crosslink type and that the ear resistance of compounds with different crosslink type are ranked the same order as tensile strength, there is surprisingly little ublished data. Some results [23] for SBR, which does not strain rystallise, are shown in Fig. 16.3 where the marked inferiority of the arbon–carbon crosslinks to polysulphidic crosslinks is apparent. The esults are expressed in terms of the tearing energy which is the work which has to be done in order to create unit area of newly torn urface [24]. The tearing energy concept has been successfully ap-lied to all aspects of strength behaviour [25–28] and makes the esults independent of testpiece shape. Its value depends on the

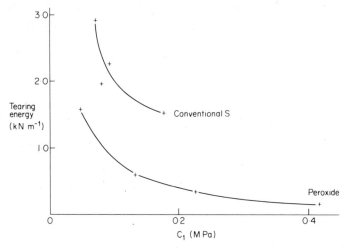

IG. 16.3. Tearing energy for unfilled styrene–butadiene rubber as a function of crosslink density assessed by means of C_1 from the Mooney–Rivlin equation [23].

externally applied forces, the flaw size, the testpiece dimensions, and
the stress–strain properties of the rubber for a crescent tear testpiece. In
contrast it is simply related to the tearing force and testpiece thickness
for the 'trouser' testpiece which is the one used to obtain the data shown
in Fig. 16.3.

Carbon black fillers increase the tear resistance considerably, particularly at low rates of tearing [29, 30], so that it seems likely that the
effect of crosslink type will be reduced in these regions.

16.3   CUT GROWTH AND FATIGUE FAILURE

Fatigue failure in rubbers is due to the growth of flaws under repeated
stressing. Flaws of the order of a few hundredths of a millimetre are
present in all rubbers but accidental surface cuts, or scratches, or
particles of foreign matter (*e.g.*, dirt), or cracks owing to ozone attack
on unsaturated rubbers, may all produce much larger flaws than those
which occur naturally. In strain crystallising rubbers, such as natural
rubber or polychloroprene, the cracks only grow when the rubber
surrounding the flaw is being subjected to an increasing tensile stress.
If the stress is held constant the growth of the flaws cease. Non-strain
crystallising rubbers do not show this behaviour and the cracks grow
all the time that a tensile stress is present. The growth of artificially
introduced cuts in rubbers has been studied [31] and the results for
various types of crosslink are shown in Fig. 16.4. The cut growth rate

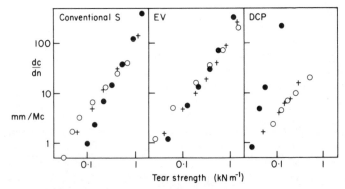

FIG. 16.4. Mechanico-oxidative cut growth characteristic of unfilled natural rubber
[31]. ○, 35 IRHD; +, 47 IRHD; ●, 54 IRHD.

is a unique function of the tearing energy for each rubber compound but the differences are relatively small except for those compounds having a high crosslink density. In these cases the crack growth rate appears to depend more strongly on the tearing energy than is the case at lower crosslink densities.

At very low tearing energies cut growth owing to the repeated deformation of the rubber ceases [32] and the value of the tearing energy, T_0, at which this occurs appears to be influenced by crosslink type [33] as shown in Table 16.10. The value of T_0 is influenced by the

TABLE 16.10
Strength properties of gum natural rubber [33]

Property	Crosslink type		
	S_x	S_1	C–C
Tearing energy, T_0 (Jm^{-2})	25	15	20
Tensile strength, TS (MPa)	26·5	23	20

presence of antioxidants and the values shown may be increased by a factor of almost two by the choice of a suitable antioxidant. The reason for the polysulphidic crosslinks giving higher values of T_0 is thought to be due to their labile nature under the action of high stresses at the tip of the cut. This allows the forces on the chains to be more uniformly distributed around the tip of the cut.

The fatigue life of a component depends amongst other things on the maximum value of the tearing energy during the deformation since the tearing energy determines the rate of crack growth. For simple testpieces, the tearing energy can be calculated, but for more complex components this is not always possible. Nevertheless the fracture mechanics approach described above has been extremely useful in identifying the factors which are important in determining the fatigue life of components. It is also useful in assessing the value of laboratory tests when comparisons between different compounds are made. For example, if tests are done at constant strain and the stress–strain properties of the compounds under investigation are different, then the results are not truly comparable. Equally, differences in stress relaxation rate will affect the fatigue lives since the maximum stress will decrease at different rates for the compounds. This latter point is extremely important and is often overlooked. It can lead to erroneous conclusions since many components operate under constant stress conditions rather than constant strain.

It has been suggested [14] that main-chain modification is more important than crosslink type in some instances as shown in Fig. 16.5. It can be seen that at low stored energies the compounds with cyclic sulphide main-chain modification show good fatigue life while those with pendent groups do not. This effect is not carried through to high values of the stored energy where the conventional vulcanisate is markedly superior to all the monosulphidic crosslinked compounds whatever their main-chain modification. The effects could be due to hysteresis which is known to affect the crack growth properties [33]. At high strains, the hysteresis arises mainly from crystallisation which is not so marked in rubbers with monosulphidic crosslinks, but at low strains hysteresis is related to internal friction which may be assessed from resilience measurements. Only one of the two compounds with good fatigue life had its resilience measured but it was the lowest of all those measured.

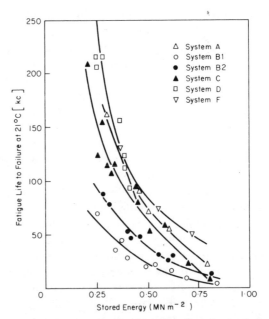

FIG. 16.5. Fatigue life of natural rubber as a function of stored energy density at maximum strain [14]. See Table 16.6 for explanation of key.

16.4   RELAXATION PROCESSES

Two types of relaxation process have been identified [34]. At normal temperatures the creep and stress relaxation are proportional to the logarithm of the time and are thought to be due to the physical rearrangement of molecules in the final approach to equilibrium after a deformation. This process has an exponential time dependence and is called a secondary, or chemical, relaxation [35]. Both types of relaxation are important in engineering applications because they may cause the deformation to become unacceptably large during the lifetime of a component.

16.4.1   Physical relaxations

STRESS RELAXATION AND CREEP

Stress relaxation and creep are intimately related and one may be calculated from the other if the stress–strain behaviour of the rubber is known [36]. The type of crosslink has an effect on the stress relaxation rate [37] which is a function of the molecular weight between crosslinks, M_c, as shown in Fig. 16.6. It can be seen that at higher degrees of crosslinking (i.e., low M_c values) the differences between crosslinks are marked but at low degrees of crosslinking three of the curves meet. The data for the urethane crosslinks seem to be consistently higher than those of the other compounds at all values of M_c. The surprising feature is that the stress relaxation rate for monosulphidic crosslinks appears to be greater than that for polysulphidic crosslinks. The difference is not large and consequently a specific experiment was carried out to investigate the effect of changing the polysulphidic crosslinks to monosulphidic crosslinks chemically, using the triphenylphosphine treatment referred to earlier. This treatment did not affect the crosslink density assessed from elastic measurements and there was little effect on the creep rate over the first two decades of time (up to 100 min) but at longer times the creep rate of the treated rubber increased noticeably probably owing to oxidation effects since the antioxidant was leached out during the treatment. This experiment suggests that polysulphidic crosslinks do not result in greater creep and stress relaxation at shorter times.

Recently [3] some doubt has been cast on the validity of earlier work since it has been shown that the rate of stress relaxation can be substantially reduced in natural rubber by the use of soluble com-

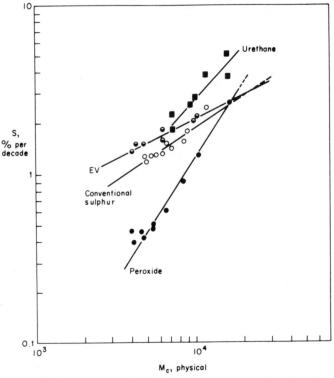

FIG. 16.6. Stress relaxation rate, S of gum natural rubber as a function of molecular weight between crosslinks, M_c. (Data from ref. 37 except for urethane.)

pounding ingredients and a purified form of natural rubber (*i.e.* deproteinised rubber, DPNR) as shown in Fig. 16.7. Data for urethane crosslinked rubber are also shown and again indicate a rather high rate of relaxation. In this context, the use of soluble compounding ingredients refers to the solubility at room temperature, since it has already been pointed out that solubility at the vulcanising temperature is essential. If the compounding ingredients are not soluble at room temperature they are liable to precipitate in the bulk of the rubber and it is this precipitate which has deleterious effects on stress relaxation and creep. It is necessary to keep the sulphur level reasonably low so that a semi-EV or EV system is required. The accelerators are carefully chosen to ensure solubility at room temperature and the stearic acid activator is replaced by a soluble zinc soap (zinc 2

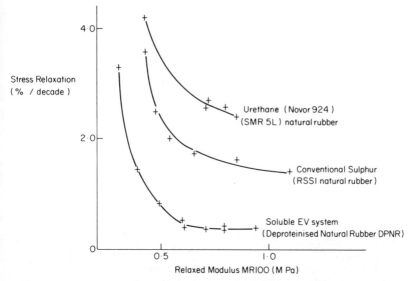

FIG. 16.7. Stress relaxation rate of gum natural rubber as a function of the relaxed modulus, MR100. (Data from ref. 3 except for urethane.)

ethylhexanoate). The stress relaxation rate of these compounds is similar to those with carbon–carbon crosslinks (*i.e.*, between 0·5 and 1 per cent per decade) so it appears that the effect of crosslink type, if any, is swamped by the presence of insoluble materials which are normally present in rubber compounds.

The presence of carbon black generally increases the stress relaxation rate [38] but the superiority of the fully soluble monosulphidic system is still observed as shown in Fig. 16.8 for two different blacks. The effect is most pronounced for a high structure SRF black (Magecol 888) but it is still apparent with a normal SRF black both for SMR5CV and for DPNR grades of rubber. In contrast, measurements on FEF-filled rubbers suggest that there is no effect of crosslink type on stress relaxation rate [38].

PERMANENT SET

Conventional 'permanent set' tests may not give a true indication of permanent set since the recovery period is too short relative to the deformation period [36]. Many compounds showing 'permanent set' in these tests would recover almost completely if sufficient time were

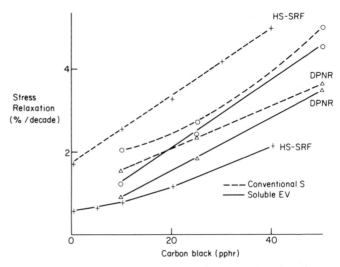

FIG. 16.8. Stress relaxation rate of filled natural rubber as a function of carbon black loading [3, 22]. Rubber SMR5CV, black SRF except where shown otherwise. DPNR, deproteinised natural rubber.

allowed. This test is therefore a disguised form of stress relaxation test.

Permanent set of a non-standard form has been used to investigate the labile nature of crosslinks [13, 39]. A dumbell is stretched to a high strain for a short period (1 min) and then allowed to recover for a long period (24 hr) so that short-term effects are eliminated. In practice, a fixed tensile stress of $150 \, \text{kg} \cdot \text{cm}^{-2}$ was used on natural rubber vulcanisates with different types of crosslinks as shown in Table 16.11. The polysulphidic crosslinked rubber shows the largest amount

TABLE 16.11
High stress permanent set in gum natural rubber[a]

	Crosslink type			
	S_x	S_1	U^b	C–C
Set (%)	9	2·5	7	1·7
Tensile strength, TS (MPa)	29	25	20	22

[a] Data from ref. 13 except for urethane.
[b] Urethane system.

of set closely followed by the urethane crosslinked rubber, while the monosulphidic and carbon–carbon crosslinked rubbers have substantially less set. These data for the urethane crosslinked rubbers are in accord with the stress relaxation data, which suggest that the crosslinks are relatively weak. The other results are in accord with the hypothesis put forward to explain the tensile strength that polysulphide crosslinks are mechanically labile at high stresses, whereas monosulphide and carbon–carbon crosslinks are not. The relatively low tensile strength of the urethane crosslinked rubber suggests that although these crosslinks are weak mechanically they do not reform in the way suggested for polysulphide crosslinks.

Compression set is a simple test which may be carried out at constant strain or constant stress. If measurements are made at room temperature it is a measure of primary relaxation processes, whereas if it is carried out at elevated temperature it is a measure of secondary relaxation processes. When tests are carried out for 24 hr at 70°C it is difficult to decide which of these two processes is dominant.

16.5 CHEMICAL RELAXATIONS

16.5.1 Stress relaxation and creep

Relaxation at high temperatures or very long times is due to oxidative scission of the main chains in the rubber and although the type of crosslink has an effect, the nature of the antioxidant is much more important [40]. Monosulphidic and carbon–carbon crosslinked rubbers showed much less creep *in vacuo* than in air and much greater creep in air after extraction of the antioxidant with hot acetone [34] in accordance with the hypothesis that oxidative scission is responsible for the creep. A conventional vulcanisate showed rather surprising behaviour in that the creep rate *in vacuo* was higher than in air. Vulcanisation for longer times which reduces the proportion of polysulphidic crosslinks did not affect the creep in air but reduced that *in vacuo* until it eventually became less than that in air. The relaxation *in vacuo* is attributed to crosslink failure and it is supposed that in air the crosslinks are more stable for reasons which are not clear.

16.5.2 Compression set

As mentioned earlier, if this test is carried out at a sufficiently high temperature it is a measure of chemical relaxation processes. For

TABLE 16.12

Compression set of black filled natural rubber [22, 41, 42][a]

Black pphr	Black type	Vulcanising system	RSSI (% set)	Compound No.	SMR 5 (% set)	Compound No.
0	–	conventional S	28	29	–	–
0	–	EV	13–19	30–33	–	–
50	HAF	conventional S	31	36	27	10
50	HAF	EV	17	37	10	12
50	HAF	U	–	–	31	15
50	FEF	conventional S	30	38	–	–
50	FEF	EV	13	39	–	–
50	Lamp	conventional S	21	40	–	–
50	Lamp	EV	13	41	–	–
50	SRF	conventional S	–		29	16
50	SRF	EV	–		10	20
50	SRF	soluble EV	–		13	19
50	SRF	peroxide	–		5	21

[a]See Tables 16.3 to 16.8 for compound formulations.

ractical purposes it is desirable that compounds should have as little
t as possible since any change represents a departure from the
itial design values. Results for different crosslinks are shown in
able 16.12. The type of carbon black does not appear to have a
ajor effect and only lampblack shows any significant difference
tween conventional sulphur compounds including a gum compound
1]. The EV systems show significantly less set in all cases while the
rethane crosslinked rubber shows a similar amount of set to a
onventional vulcanisate. A surprising feature is that the soluble EV
ystem is slightly worse than the normal EV, although it is doubtful
hether differences of this magnitude are significant. The best sul-
hur system (not shown in the table) is a soluble EV vulcanisate using
PNR and the set of this compound (8 per cent) is close to that of a
eroxide vulcanisate [22].

The results are difficult to interpret in terms of other relaxation
rocesses since the black level and type do not affect compression set
ery much but they do affect stress relaxation and creep behaviour.
evertheless, the effect of crosslink type follows the pattern obser-
ed in stress relaxation and creep measurements in that the con-
entional system is not as good as EV vulcanisate which is inferior to
carbon–carbon crosslinked rubber. A soluble EV system with
PNR may be almost as good as a peroxide system. The urethane
rosslinked rubbers appear to be similar to conventional vulcanisates.

16.6 RESILIENCE AND HEAT BUILD-UP

he dynamic properties of compounds are important in many spring-
g applications. It is usually necessary to test compounds under
imilar deformations and at similar rates as occur in service, preferably
sing the actual component in a suitable test rig. A useful indication
f dynamic properties can be obtained very quickly from resilience
easurements or within a period of one or two hours using a
oodrich flexometer or similar machine. It is not usually possible to
btain precise correlations with data from these machines and service
erformance, but they are probably satisfactory in giving a reasonable
idication of the likely behaviour.

Both crosslink type and crosslink density affect the resilience of
ulcanisates as shown in Fig. 16.9. The effect of crosslink density is
o be expected [14] since more crosslinks reduce the deformation

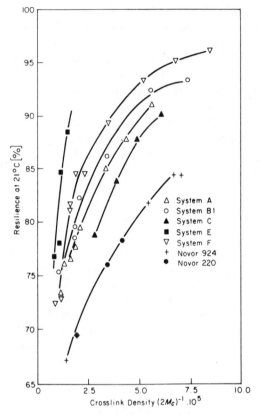

FIG. 16.9. Resilience of gum natural rubber vulcanisates measured on a Dunlo tripsometer as a function of crosslink density. (Data from ref. 14 except for urethane See Table 16.6 for explanation of key.

arising from the impact of the indentor and since the energy loss due to internal friction in the rubber a reduction in the deformatio will increase resilience. The influence of crosslink type is rathe surprising. At a particular crosslink density the polysulphidic cros linked material gives the highest resilience although the range c crosslink density is rather small. Next best is the conventional syste containing about 70 per cent polysulphidic crosslinks. The results fe the monosulphidic and carbon–carbon crosslinked rubbers all le below those of the conventional material. The results for the urethan crosslinked rubber appears to be significantly below the rest and it tempting to suggest that this may be due to main-chain modificatior which have increased the glass transition temperature. Evidence to support this hypothesis comes from the next lowest curve which shows data for a monosulphidic crosslinked rubber containing more main-chain modifications (assessed chemically) than any of the other compounds. However, all the data do not support the view that main-chain modification is the sole factor since the conventional vulcanisate contains more main-chain modification than the monosulphidic crosslinked rubber just below it. Also the carbon–carbon crosslinked rubber has a rather low resilience considering that it is thought to have very little main-chain modification.

The effect of crosslink type is also apparent in black filled compounds containing 50 HAF pphr as shown [42] in Table 16.13. Here again the EV systems have lower resilience than the conventional but the semi-EV shows the highest resilience of all. Before seeking an explanation of this effect it would be necessary to find out whether it is really due to crosslink type, in which case it should also occur in unfilled compounds. The urethane crosslinked materials again show significantly lower resilience than the other compounds. Measurements on compounds with a different black (50 SRF pphr) at a higher temperature (50°C) show much higher resiliences for all the compounds as expected. The results are also shown in Table 16.13 where it can be seen that the peroxide vulcanisate is now very similar to the conventional system and much of the deficiency of the EV system is removed by using soluble compounding ingredients.

The black filled compounds have also been tested on a Goodrich flexometer which gives a cumulative assessment of energy losses as shown by the temperature rise which results from repeated compressive deformations of the sample (which is in the form of a cylinder). The results are shown in Table 16.13 where it can be seen that the temperature rise is not necessarily related to the resilience. In the conventional test, measurements are carried out at room temperature but some data in these tables were obtained at an elevated ambient temperature of 120°C. The room temperature results in Table 16.13 show the conventional vulcanisate to have a lower temperature rise and more set than the EV systems. The urethane crosslinked compounds show much higher temperature rises and much greater set. When measurements are carried out at the elevated temperature a different picture emerges as shown in Fig. 16.10. The conventional vulcanisate fails and the EV system together with the urethane system are quite satisfactory. The soluble compounding ingredients

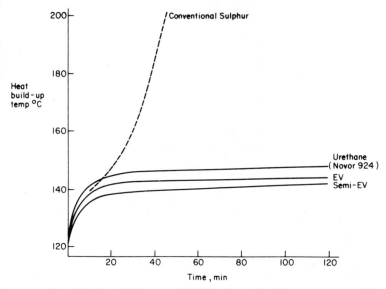

FIG. 16.10. Goodrich heat build-up measured at elevated ambient temperature (120°C) for natural rubber [42].

do not appear to be particularly beneficial on these tests. The reason for the poor performance of the conventional vulcanisate is reversion at the higher temperature.

16.7   AGEING BEHAVIOUR

The ageing behaviour of vulcanisates is normally assessed by exposing testpieces to high temperatures in an air oven for prescribed periods of time. Any physical property of interest can be measured after ageing and the results are usually expressed either as the percentage retained after ageing or as the percentage change as a result of the ageing. The physical properties most commonly measured are hardness, modulus and tensile strength but any property which is considered appropriate for the particular application may be used. The results for unfilled rubbers, shown in Table 16.14, indicate the marked superiority of EV systems (monosulphidic crosslinks) over conventional systems (predominantly polysulphidic crosslinks). The urethane crosslinked rubbers are not quite as good as the EV compounds but are substantially better than the conventional system.

TABLE 16.14

Retention of tensile strength (%) of aged natural rubber[a]

Black				Days aged at:					
				100°C				125°C	
pphr	type	Vulcanising system	Compound No.	3	5	7	14	3	5
0	–	conventional S	29	–	–	6	B[b]	–	–
0	–	EV	30–33	–	–	48–77	39–68	–	–
0	–	urethane	1–7	93	–	40–54	10–24	B	–
50	HAF	conventional S	10, 29	–	–	25	15	21	–
50	HAF	EV	10, 29	–	–	78	68	50	–
50	HAF	urethane	15	–	–	52	–	6	–
50	FEF	conventional S	38	–	–	24	18	–	–
50	FEF	EV	39	–	–	79	65	–	–
50	Lamp	conventional S	40	–	–	31	21	–	–
50	Lamp	EV	41	–	–	74	67	–	–
50	SRF	conventional S	16	–	35	–	–	–	B
50	SRF	EV	20	–	85	–	–	–	31
50	SRF	soluble EV	19	–	86	–	–	–	20
50	SRF	peroxide	21	–	58	–	–	–	48

[a] Data from refs 22, 41 and 42 except for urethane gum. See Tables 16.3 to 16.8 for compound formulations.
[b] B, brittle.

This behaviour is carried through into HAF-black-filled compounds s also shown in Table 16.14. The type of black does not seem to nfluence the effect of crosslink type greatly although the black filled ompounds are generally better than gum compounds, probably wing to the use of a more effective antioxidant which is staining and herefore not normally used in gum compounds. At higher temeratures, *e.g.*, 125°C, and longer times, *e.g.*, 5 days, the advantage of he peroxide vulcanisate is apparent, retaining nearly 50 per cent of its riginal tensile strength compared with 20 per cent for a soluble EV nd 31 per cent for a normal EV.

The pattern of behaviour that emerges is relatively simple. The ageing behaviour of conventional compounds containing redominantly polysulphide crosslinks is much worse than either urethane crosslinked or monosulphide crosslinked rubbers. Although black filled compounds generally show better ageing behaviour than unfilled compounds, the effect of crosslink type is still apparent. Carbon–carbon crosslinks are particularly beneficial at higher temperatures, *e.g.*, 125°C, but not so good at lower temperatures relative to the other crosslink types.

The fatigue behaviour of compounds with different crosslink types is also affected by ageing as shown in Fig. 16.11. In the unaged

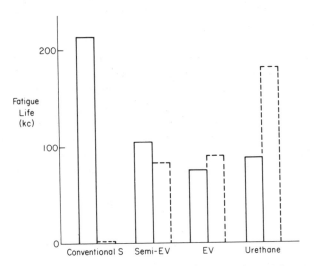

FIG. 16.11. Comparative ring fatigue behaviour of aged and unaged natural rubber vulcanisates [42].———, Unaged;-----, aged for 7 days at 100°C.

samples the polysulphide crosslinked material is best with monosu
phide crosslinked and urethane crosslinked rubbers having less tha
one-half the fatigue life as measured on a ring fatigue machine. Afte
ageing for 7 days at 100°C, the life of the conventional system ha
been reduced almost to zero, whereas the monosulphide crosslinke
rubber has hardly changed and the urethane crosslinked rubber ha
more than doubled its fatigue life. The behaviour of the urethan
crosslinked rubber is rather surprising and it is not clear whether it
a genuine effect or a consequence of changes in other physic
properties (*e.g.*, modulus or set). Nevertheless, there is no doubt tha
both urethane and monosulphidic crosslinks give substantially im
proved performance over conventional vulcanisates.

16.8 CONCLUSION

It is evident when considering the properties described earlier th
there is no crosslink type which is best for all purposes. The co
ventional sulphur vulcanisate with 70 per cent polysulphidic cros
links has the best tensile strength, tear resistance and resilience b
poor stress relaxation and ageing behaviour. The monosulphid
crosslinked rubber gives much better stress relaxation only if th
compounding ingredients are soluble, but the ageing behaviour
considerably improved at the expense of lower tensile strength ar
resilience. The urethane crosslinked rubber has good agei
behaviour but there is evidence that at least some of the crosslin
break easily (high stress relaxation and tension set) and do n
re-form (low tensile strength). The presence of weak or tempora
crosslinks is rather surprising since the urethane crosslinks a
expected to be very stable from a chemical point of view. In additio
the low resilience suggests that there may be some main-cha
modification. The urethane system appears to give a higher crossli
density in black filled compounds than in gum compounds and t
curing characteristic is very flat so that long cure times are possib
without a deterioration of the physical properties through overcure
the rubber near to the surface. The carbon–carbon crosslinked rubb
has very low set and stress relaxation but poor tensile strength ar
tear resistance. The ageing behaviour at high temperatures (*e.
125°C) is better than that of any of the other systems. The influen

of crosslink type is still apparent in black filled compounds although the differences are generally smaller.

The choice of crosslink type may be quite difficult in some cases requiring considerable expertise in balancing the various conflicting requirements of the desired physical properties and also the curing characteristics in the case of thick components so that a satisfactory performance is achieved.

ACKNOWLEDGEMENTS

I am indebted to Dr C. S. L. Baker at MRPRA for providing the urethane crosslinked rubbers and to my former colleagues at the same establishment for helpful discussions in the preparation of this paper. I would particularly like to thank Professor A. G. Thomas and Dr M. R. Porter for their generous advice.

REFERENCES

1. CHARLES GOODYEAR. (1855). *Gum Elastic and its Varieties.* New Haven.
2. THOMAS HANCOCK. (1920). Personal narrative of the origin and progress of caoutchouc in England. J. L. Hancock Ltd.
3. J. F. SMITH. (1974). *Rubber in Engineering Conference*, Kuala Lumpur. MRPRA reprint No. 764.
4. D. J. ELLIOT and B. K. TIDD. *Prog. Rubb. Tech.*, 1974, **34**, 83.
5. B. SAVILLE and A. A. WATSON. *Rubb. Rev.*, 1967, **40**, 100.
6. D. S. CAMPBELL. *J. Appl. Poly. Sci.*, 1969, **13**, 1201.
7. D. S. CAMPBELL. *J. Appl. Poly. Sci.* 1970, **14**, 1409.
8. D. S. CAMPBELL. *J. Appl. Poly. Sci.*, 1971, **15**, 2661.
9. M. BRADEN and W. P. FLETCHER. *Trans. Instn Rubb. Ind.*, 1955, **31**, 155.
10. M. BRADEN, W. P. FLETCHER and G. P. McSWEENEY. *Trans. Instn Rubb Ind.*, 1954, **30**, 44.
11. D. S. PEARSON and G. G. A. BÖHM. *Rubb. Chem. Tech.*, 1972, **45**, 193.
12. C. S. L. BAKER, D. BARNARD and M. PORTER. *Kaut. und Gummi Kunst.*, 1973, **26**, 540.
13. L. MULLINS. (1967). In: *Proceedings of Polymer Structure and Mechanical Properties.* US Army Natick Labs, Natick, Mass.
14. G. M. BRISTOW and R. F. TILLER. *Kaut. und Gummi Kunst.*, 1970, **23**, 55.

15. J. M. GOPPEL and J. J. ARLMAN. *Appl. Sci. Res.*, 1949, **1**, 462.
16. G. GEE. *J. Poly. Sci.*, 1947, **2**, 451.
17. P. J. FLORY, N. RABJOHN and M. C. SCHAFFER. *J. Poly. Sci.*, 1949, **4**, 435.
18. J. LAL and K. W. SCOTT. *J. Poly. Sci.*, C(9), 1965, 113.
19. L. BATEMAN (Ed). (1963). *The Chemistry and Physics of Rubber-like Substances*, p. 730. London: Maclaren.
20. A. V. TOBOLSKY and P. F. LYONS. *J. Poly. Sci.*, A2, 1968, **6**, 1561.
21. S. D. GEHMAN. *Rubb. Chem. Tech.*, 1969, **42**, 659.
22. D. J. ELLIOTT. 'Compounding natural rubber for engineering applications.' *N. R. Tech. Bulletin*, MRPRA.
23. A. G. THOMAS, unpublished.
24. R. S. RIVLIN and A. G. THOMAS. *J. Poly. Sci.*, 1953, **10**, 291.
25. G. J. LAKE, P. B. LINDLEY and A. G. THOMAS. (1969). In: *Proceedings of the 2nd International Conference of Fracture*, Brighton, UK, 1969, p. 493.
26. G. J. LAKE and A. G. THOMAS. *Kaut. und Gummi Kunst.*, 1967, **20**, 211.
27. A. G. THOMAS. (1966). In: *Conference Proceedings on Physical Basis of Yield and Fracture*, Oxford, UK, 1966, p. 134.
28. A. N. GENT, P. B. LINDLEY and A. G. THOMAS. *J. Appl. Poly. Sci.*, 1964, **8**, 455.
29. H. W. GREENSMITH. *J. Poly. Sci.*, 1956, **21**, 175.
30. L. MULLINS. *Trans. Instn Rubb. Ind.*, 1956, **32**, 231.
31. G. J. LAKE and P. B. LINDLEY. *Rubb. J.*, 1964, **146**, 24, 30.
32. G. J. LAKE and P. B. LINDLEY. *J. Appl. Poly. Sci.*, 1965, **9**, 1233.
33. G. J. LAKE and A. G. THOMAS. *Proc. Roy. Soc. (A)*, 1967, **300**, 108.
34. A. N. GENT. *J. Appl. Poly. Sci.*, 1962, **6**, 442.
35. L. BATEMAN (Ed.) (1963). *The Chemistry and Physics of Rubber-like Substances*, p. 215. London: Maclaren.
36. A. N. GENT. *J. Appl. Poly. Sci.*, 1962, **6**, 433.
37. E. D. FARLIE. *J. Appl. Poly. Sci.*, 1970, **14**, 1127.
38. M. J. GREGORY. *N. R. Tech.*, 1977, **8**, 1.
39. A. G. THOMAS. *J. Poly. Sci. Symp.*, **48**, 1974, 145.
40. L. BATEMAN (Ed.) (1963). *The Chemistry and Physics of Rubber-like Substances*, p. 738. London: Maclaren.
41. R. M. RUSSELL, T. D. SKINNER and A. A. WATSON. *Rubb. Chem. Tech.*, 1969, **42**, 418.
42. C. S. L. BAKER. 'Vulcanisation with urethane reagent.' *N. R. Tech. Bulletin*, MRPRA.

Chapter 17

THE EXPECTED SERVICE LIFE OF
STRUCTURAL ADHESIVES

W. C. WAKE, K. W. ALLEN and S. M. DEAN

17.1 THE ENDURANCE LIMIT OF ADHESIVE JOINTS

It is a major problem in the design of adhesively bonded structures to relate the design stress to the strength of overlap shear joints obtained by the usual test procedures on the small scale. Modern aircraft, such as the wide-bodied 'Jumbo' jet, are expected to fly safely for 90 000 hr, distributed over 30 years and including landing and take-off in widely differing climates. During flying and particularly in the manoeuvres of take-off and landing, the adhesive joints are subjected to cyclical stressing but, even when stationary, the joints may be subjected to a static, continuous load. There are many facets which give the problem of designing aircraft their unique degree of complexity. In an attempt to simplify one may pose the fundamental question of whether there exists a load which an adhesive joint will sustain indefinitely under all conditions.

The approach to this problem which has been used by previous workers had also looked back to the treatment of high-speed cyclic loading tests used by metallurgists in studying the fatigue of steel and other alloys. This involves the use of the hyperbola-like curves obtained by plotting the maximum stress of a stress cycle against the number of cycles sustained before failure. Such tests make heavy demands on machine investment and time. They are additionally expensive because, as will be realised on reflection, the test machine as well as the test specimen is subjected to fatigue and has a limited life. Nevertheless, lap-shear adhesive joints have been subjected to prolonged cyclical testing and shown to be superior to riveted joints

311

[1]. On the assumption that the so-called S–N or Wöhler curves were
hyperbolic, Prot [2] argued that determination of tensile strength by
tests in which the rate of stressing increased linearly during the
progress of the test would enable the S–N curve to be transected at
several points and hence defined and constructed without actually
undertaking the series of long-term cyclical loading tests otherwise
required. One of the asymptotes of the hyperbola is the *endurance
limit*, a load below which failure was not expected. The details of the
derivation of S–N curve from data obtained by constant rate of
loading are not of concern in the present paper. Suffice it that
Loveless, Deeley and Swanson [3], developed from Prot's method an
expression which was used by Lewis and his colleagues [4] to obtain
an endurance limit for lap shear adhesive bonds with a variety of
adhesives. They then attempted to validate the figures they obtained.
   It would be an important advance if it could be accepted that there
was an endurance limit even if the figure obtained referred, as would
be expected, to joints in which the adhesive was maintained free from
chemical degradation by way of oxidation or hydrolysis and the mode
of failure remained cohesive within the adhesive without any moisture contributing to interfacial failure. Lewis and his colleagues
showed the ratio of the endurance limit to the short-time tensile
strength to be a constant independent of adhesive modulus, relatively
insensitive to joint geometry and temperature of test. They noted
however, that the predicated hyperbolic relation demands that the
time-to-break under a constant load is inversely proportional to the
difference between the applied stress and the endurance limit but that
their results were better fitted by a logarithmic relation.
   The present work related back to the original concept of dynamic
rather than static behaviour and to the argument that successive
cycles of stress caused the growth of cracks from existing flaws. If
this represented a reasonable model then cycling at stresses below the
endurance limit should not result in either failure or loss in tensile
strength whereas cycling above this limit should lead to a progressive
decrease of joint strength and sporadic failures.

17.2   THE INFLUENCE OF CYCLIC STRESS ON
THE STRENGTH OF JOINTS

17.2.1   Determination of endurance limit

Batches of lap-shear adhesive joints were manufactured from pairs of
aluminium plates 9×4 in. The aluminium plates were prepared by

vapour degreasing in trichloroethylene followed by chrome–sulphuric treatment in accordance with aircraft specification DTD 919B. The adhesive was a polyvinyl formal–phenolic adhesive Redux 775* applied in film form and cured by heating to 163°C for 30 min under a pressure of 100 p.s.i., two sets of joints, four plates at one time. Each joined pair was cut into nine joints each $7 \cdot 5 \times 1$ in. with a 0·5-in. overlap. The two end joints which contained locating holes and the centre one where this contained a thermocouple were discarded. All joints were checked for symmetry, overlap consistency and glue line thickness before acceptance for testing.

Batches of these joints were tested to failure in a Mayes machine (model ESHM 250) at different rates of *stress* increase. In these tests the time to failure was also measured carefully. Average values of these data are given in Table 17.1 together with the necessary derived functions.

TABLE 17.1
Experimental and derived parameters for determination of endurance limit

Stress at break (kN)	α (kN·min^{-1})	t_b (min)	t_b^2 (min^2)	αt_b^2 (kN·min)
10·67	0·223	48·70	2371·690	528·89
10·88	0·445	24·41	595·848	265·15
11·05	2·225	4·91	24·108	53·64
11·13	4·450	2·49	6·200	27·59
11·59	22·250	0·54	0·292	6·49
11·84	44·500	0·27	0·073	3·24
12·05	219·981	0·08	0·006	1·41

Loveless *et al.* deduced the equation

$$\alpha t_b^2/2 = K - (S_0 - \text{EL})t_b \qquad (17.1)$$

where t_b = time to break, S_0 = a pre-load term (discussed below), α = rate of increase in stress, K = a material constant and EL = endurance limit. S_0 was an adjustable parameter with the dimensions of stress to allow for some stress annealing to occur. A plot of αt_b^2

---

*Redux is a registered trade name, the property of Ciba-Geigy Ltd, Duxford, Cambs.

against t_b has a positive slope showing that $EL > S_0$ and enables $(EL - S_0)$ to be estimated from the data given in Table 17.1 as 5.44 kN and the intercept gives the constant K as 0.269 kN·min^{-1}.

The assumption is now made that if S_0, the stress annealing term, has a real value greater than zero then it is small in relation to the endurance limit.

The stress at break increases with increasing rate of loading but over the range used, 1000-fold, the increase is only 13 per cent. Reproducibility of the figures was such that replication between 8- and 16-fold showed standard errors for their means of about 0.1 kN. Thus the endurance limit is between 44 and 50 per cent of the ultimate strength.

17.2.2 Effect of cycling on joint strength

From a batch of joints, six were chosen randomly and their tensile strength determined in the Mayes machine with a rate of loading of 4.45 kN·min^{-1}. The remaining joints were used in cycling experiments before similarly subjecting them to tensile test. The stress cycling was achieved automatically in the same machine. It was through a fixed *strain* cycle determined by limit switches and corresponding to a maximum stress of either 60 or 35 per cent of the ultimate tensile strength thus well clear, both above and below, of the range for the endurance limit. The lower end of the stress cycle was kept above zero, actually at 5 per cent of the ultimate tensile strength, to ensure that at no time any region of the joint experienced a compressive stress.

All these experiments were performed at laboratory temperature without temperature control.

The results of these cycling experiments are shown in Table 17.2.

Preliminary results established that at 15 cycles per minute (c.p.m.) no significant temperature rise was recorded by thermocouples implanted in the glue line but the time taken on the testing machine to produce measureable change at 0.5 c.p.m. was obviously too long. Experiments were therefore limited to cycling at 10 and 15 c.p.m.

For the results reported in Table 17.2, 78 joints were used but nine failed whilst undergoing cycling. All failures occurred in the group which was cycled above the endurance limit. Table 17.2 is based on the median of the number of tensile results (recorded in parentheses in the table), the median referring only to testpieces surviving the test with no account being taken of failed specimens. No cycling below

TABLE 17.2

Median joint strengths after periods of cyclical stress

No. of cycles	Post-cycling tensile strength of joint as percentage of initial strength	
	10 cycle·min^{-1}	15 cycle·min^{-1}
Above EL[a]		
75		96(6)[b]
150	87(7)	93(6)
200		88(6)
250	91(5)	
400	84(13)	
Below EL[a]		
75		103(3)
150	100(4)	
200		101(3)
400	100(8)	99(4)
1,000	99(2)	98(2)

[a]*EL* = Endurance limit.
[b] Figures in parentheses indicate the number of joints involved.

he endurance limit has produced significant reductions in tensile strength, whereas cycling above the endurance limit has produced significant reductions in strength and moreover, has led to a proportion of failed joints.

17.2.3 Joints failing during cycling

Table 17.3 shows the cumulative occurrence of failed joints against the number of cycles at which successive failures occurred.

The number of joints at risk at the start of the experiment consisted of those that survived to be tested and those which failed, a total of 52 which were subjected to stress cycling above the endurance limit. If cumulative percentage is plotted against the log(number of cycles survived) as in Fig. 17.1, a more linear representation of the data is obtained than with an arithmetic plot, indicating failure times to be distributed as a log-normal rather than a normal distribution with a median value (the mean of the log-normal distribution) at about 1500

TABLE 17.3
*Occurrence of joint failure as function of
number of cycles*

No. of cycles	Total joints failed	Joints failed as percentage of those under test
80	1	1·9
145	2	3·8
180	3	5·8
214	4	7·7
240	5	9·6
315	6	11·5
396	7	13·5
398	8	15·4
435	9	17·3

cycles, *i.e.* 50 per cent of the joints would be expected to fail before 1500 cycles is exceeded.

17.3   EFFECT OF STATIC STRESS

An attempt was made to determine the time to break at loads corresponding to 75–90 per cent of the breaking load as normally determined in a tensile testing machine. The joints were fastened with the usual grips, the machine started and then stopped when the load reached the required figure. In this way the joints were all loaded at the same rate of loading and were free from shock loading. The scatter of the results obtained rendered them useless and recourse was had to an examination of data previously obtained [5] though with a different purpose originally in view.

Joints made with Redux 775 as the adhesive, when subject to a static load show behaviour which can be regarded as occurring in separate stages. These stages comprise: (i) an initial, unmeasured, elastic deformation; (ii) a period during which no deformational response to stress occurs, known as the delay time which varies with stress and temperature; (iii) a period during which steady logarithmic creep occurs at rates determined by stress and temperature; and (iv) a period of accelerating creep terminating in tensile rupture.

The interest at the time the work was undertaken resided in the

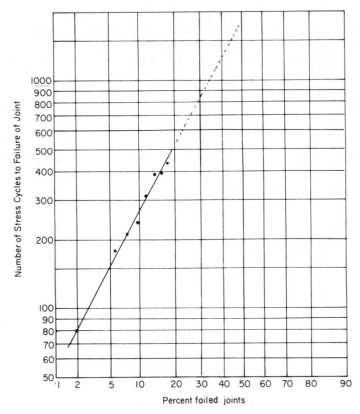

FIG. 17.1. Probability of joint failure on cyclical stressing as a function of the number of cycles.

delay time before the initiation of creep and the logarithmic creep which follows this. However, the automatic recording necessarily used in these experiments also recorded the commencement of accelerating creep and the time of tensile rupture where this occurred before the experiment was terminated. It is thus possible to obtain time to failure at known loads and temperatures. By the same argument as was used with delay times [5], and indeed originally used in connection with the failure of textile fibres [6], the time to failure (t_f) under constant load can be expressed as

$$t_f \simeq B/T \exp\left[\frac{\Delta F - \alpha S}{RT}\right] \qquad (17.2)$$

where B is a constant, ΔF is the energy of activation of the stress biassed flow process and α the volume involved in this process, S is the stress and RT has its usual significance. This expression allows the relation between time to failure and stress to be examined in the form

$$\ln t_f = \ln B/T + \Delta F/RT - \alpha S/RT \qquad (17.3)$$

or

$$\log t_f = A - mS \qquad (17.4)$$

Figure 17.2 shows that data read from Shanahan's recordings [5a] are very reasonably fitted by a relation of this form. However, the range of validity of this equation is limited by the assumptions made in it.

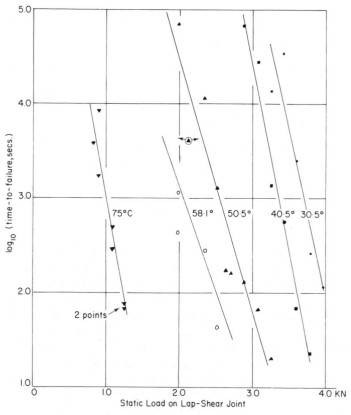

FIG. 17.2. Time to failure under static load. (After Shanahan [5a].)

derivation and the approximations in its simplification. Among these is the assumption that the stress involved in the bias of the potential energy barrier for the flow process leading to rupture is very small compared with the imposed stress. This biasing stress may therefore be regarded as equivalent to the endurance limit. If so, it implies a very much smaller endurance limit than is deduced by the modified Prot procedure or implied by the behaviour of the cyclically stressed joints described in Section 17.2.2. There is no indication that the data depart from linearity within the load range used. Unfortunately, the short-term breaking loads of Shanahan's joints were not measured and hence there is no indication of the actual magnitude which might be expected for a Prot-type endurance limit. However, it could be not more than one-half of the lowest loads used for the joints at 30·5°C. It is obvious that extrapolation to a load of this magnitude would give a lifetime of 10^{10} s—about 300 years. The data are, therefore, not inconsistent with the existence of an endurance limit.

17.4   RELATION BETWEEN DECLINE IN STRENGTH AND FAILURE

One of the more disturbing aspects of the testing of adhesive joints is the occasional occurrence of joints of exceptionally low strength or of failures in exceptionally short times. The present series of experiments suggests that 1 per cent of joints will fail under cyclical stress when the average decline in strength due to the stressing is only about 2 per cent. Normal tensile testing of batches of joints does not show a distribution of strengths as wide as these results indicate and suspicion must therefore be levelled at the test procedure. Some uncontrolled variables must exist, perhaps in specimen placement or stress to which this mode of testing is particularly sensitive. The fall in strength required is not, of course, a fall to zero but one to a value around that of the maximum stress applied during the stress cycle. It would be interesting to relate the loss in strength to the energy dissipated during the cycle and to know if this increases before failure in those joints which fail prematurely. Until the spread in the number of cycles to failure can be reduced no reasonable relation between strength and failure rate can be expected. Coleman and Knox [6] draw attention to the fact that with textile fibres cyclically loaded, it is the time to failure which is important, independent of the number of cycles within that time, whereas with metals it is the number of cycles independent of the frequency. These differences imply different mechanisms, the one a rate process and the other a crack-growth process. The present evidence for adhesive joints may be summarised by stating that a rate process can account for statically loaded joints except that it is difficult to account for an endurance limit as high as dynamic methods suggest. More evidence is required from more closely controlled experiments before it can be decided if cyclical loading leads to failure by the same type of process as static loading.

There is an additional difficulty in applying the results of such experiments to the interpretation of long-term exposure trials. Under isothermal laboratory conditions, chemical change in the adhesive can be eliminated in the time-scale involved. In service, the chemical nature of an adhesive after one year's exposure cannot be assumed to be unchanged; the adhesive at the interface or the oxide surface of the substrate metal may have changed profoundly.

In the discussion of failure and strength decline in laboratory testing, whether from continuous static loading or cyclical stressing, it may be assumed to occur within the adhesive but in exposure trials interfacial or mixed failure become progressively more common as the trial proceeds.

17.5 CONCLUSION

Although somewhat outside the main theme of this book, this paper shows that strength and fatigue studies of adhesive joints have their part to play if the expected service life of adhesive joints is to be properly evaluated. Work in association with Dr Payne over many years has been a pointer to the value of cyclical dynamic experiments. Nor is this the only lesson which has been drawn from his work, for two-phase structure is also an important aspect of many of the more successful adhesives and, this too, was a field to which he made great contributions.

ACKNOWLEDGEMENTS

The work described in this paper was supported by the Materials, Quality, Assurance Directorate (MQAD) of the Ministry of Defence. The work is published with the permission of the Ministry. The authors gladly acknowledge the use of testing facilities at MQAD, Woolwich Arsenal and the value of discussions with Ministry Staff, especially A. O. van Raalte and Dr L. Brett.

REFERENCES

1. A. MATTING and U. DRAUGELATES. *Adhäsion*, 1968, **11**, 5–22, 110–132, 161–176.
2. E. M. PROT. *Rev. de Metallurgie*, 1948, **45**(12). English translation by E. G. WARD, *WADC Tech. Rep.*, 1952, 52–148.
3. H. W. LOVELESS, C. W. DEELEY and D. L. SWANSON. *SPE Trans.*, 1962, 2(2), 126.
4. A. F. LEWIS, R. A. KINMONTH and R. P. KREAHLING. *J. Adhesion*, 1972, 3(3), 249.
5. M. E. R. SHANAHAN. (a) Thesis, The City University, 1974. (b) K. W. ALLEN and M. E. R. SHANAHAN. *J. Adhesion*, 1975, 7(3), 161.
6. B. D. COLEMAN and A. G. KNOX. *Textile Res. J.*, 1957, 393.
7. British Intelligence Objectives Sub-Committee (1946). (BIOS) Final Report 791. London: HMSO.
8. A. R. PAYNE, 1956, 'Non-linearity in the dynamic properties of rubber.' In: *Proceedings 3rd Rubber Technology Conference 1954* (Ed. T. H. Messenger), pp. 413–36. Cambridge: Heffer.

HISTORICAL NOTE

One of the more peculiar aftermaths of the war was the acquisition by the Research Association of British Rubber Manufacturers (RABRM), now RAPRA, of a piece of equipment for testing the dynamic behaviour of rubber in compression from the rubber laboratories of the large German combine I. G. Farben. This equipment was designed and used by Dr. H. Roelig with whom Dr. L. Mullins and I had had extensive conversations early in 1945 [7]. Through the good offices of the late G. Hammond of the Ministry of Supply, Roelig's apparatus became available to RABRM staff for experimental work. Just after the machine was finally installed and working Payne joined us at RABRM in 1952 as a graduate in physics from Durham University.

I remember going with Bob Payne and R. H. Norman of RAPRA across to the National Physical Laboratory at Teddington to discuss peculiarities of the Roelig machine with physicists and engineers using eccentric rotating load machines in fatigue studies on metals. There were two major results of this work. The one was Bob Payne's first paper [8], an important one, given at the Third Rubber Technology Conference in 1954 and the second his interest in dynamic behaviour and the structural effects discovered with the aid of cyclical stressing of polymers. This interest persisted for the remaining 22 years of his life, coloured his approach to polymer physics and enabled him quickly to make important contributions to ideas about the structure of polyurethanes and the new poromeric materials when he became Director, in his maturity, to the Shoe and Allied Trades Research Association.

In presenting a paper to his memory it is natural therefore that I have chosen to write about dynamic properties for I and my colleagues at The City University have found the use of a cyclically acting stress useful in assessing the life behaviour of adhesive joints.

Chapter 18

THE URETHANE CONTRIBUTION TO STRENGTH AND STRESS RELAXATION IN ELASTOMERS

D. C. HARGET and C. HEPBURN

18.1 INTRODUCTION

The term 'polyurethane' is conveniently used to describe a wide variety of polymeric materials, including foams (flexible and rigid), elastomers, surface coatings and adhesives. During early development of these commercially important polymers, the predominant chemical group present in the chain was the urethane group

$$
\left[\begin{array}{c} N-C-O \\ | \quad \| \\ H \quad O \end{array}\right]
$$

However, with the rapid expansion of this class of materials over the past 30 years, the polyurethanes now include those polymers which contain a significant number of urethane groups, together with a variety of other structurally important groups, *e.g.*, ester, ether and urea groups.

18.1.1 Structure in segmented polyurethane elastomers

Most commercial types of polyurethane elastomers derive their high level of strength and flexibility from their segmented polymer structure, whose usual route of chemical formation is illustrated in Fig. 18.1. The relationships between chemical and morphological structure are crucially important in all urethanes dictating their engineering uses. Chemical composition has been well covered in other publications [1] and hence morphological considerations will now be discussed.

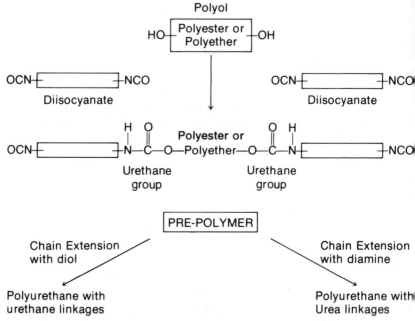

FIG. 18.1. Generalised polyurethane chemical structure.

18.2 MORPHOLOGICAL STRUCTURE

In recent years, much work has been carried out to elucidate the extent of microphase separation and the structure of the resulting components in segmented polyurethane elastomers [2, 3]. In particular, thermal analysis [4–13] and X-ray scattering investigations [7, 9, 14–20] have been found to show evidence for a two-phase structure. It is now generally accepted that the hard segments separate to form discrete domains in a matrix of soft segments. The rigid domains formed act both as tie down points (being chemically linked to the soft matrix) and as filler particles, reinforcing the soft segment matrix.

Bonart [14] has investigated the structure of segmented polyurethane elastomers using X-ray diffraction techniques. Polyether (polyoxytetramethylene) and polyester (a co-polyester) based elastomers incorporating diphenylmethane diisocyanate (MDI) and extended with ethylene diamine or hydrazine, were chosen for investigation. In the relaxed state, spatially separated hard and soft segments exist in the material. The hard segments are held together in discrete domains through the action of Van der Waals' forces and hydrogen-bonded interactions. On extension, soft-segment polymer chains undergo various extents of stretching and disentanglement, causing the rigid domains to lie in a disoriented manner transverse to the direction of stretching. In the case of polyether-based elastomers, extension beyond approximately 150 per cent gives rise to marked elongation crystallisation of the soft segments (Fig. 18.2).

Polyester-based elastomers showed a lower level of crystalline order, presumably owing to the more irregular chemical structure present. As the elastomer is further elongated up to 500 per cent extension, the orientation of soft segments improves only to a small extent, while the hard segments turn themselves with their longitudinal axes into the direction of elongation. This effect is explained in terms of force strands, *i.e.*, maximally loaded chains in the soft segment which oppose any further extension. Further extension therefore requires sliding processes between hard segments to take place, resulting in the formation of new force strands. At 500 per cent elongation, this restructuring process is virtually complete. Treatment of the extended sample with warm water (80°C) for 30 min, accelerates the process and also gives rise to an even distribution of forces amongst soft-segment chains, resulting in a loss of elongation crystallisation in this region (Fig. 18.3). Hard segments are now fully orientated in the direction of elongation, and show relatively intense reflexes in the X-ray photographs. On relaxation of the material, soft segments disorientate almost completely, while the hard segments tend to remain in the orientated manner described. This model of molecular organisation and restructuring on extension, provides a possible explanation of stress-softening phenomena and high hysteresis in polyurethane elastomers.

The factors contributing to the degree of crystallinity and domain formation in segmented polyurethane elastomers have received a great deal of attention over recent years. The effects of hydrogen-bonded interactions have been shown to be of particular importance in this context. Infra-red spectroscopic studies [21–28] have demonstrated the presence of this type of interaction. It appears that almost all NH-groups in segmented polyurethane elastomers are hydrogen bonded [22]. Hydrogen-bonded interactions between urethane groups or urea groups contribute to hard segment domain formation. There is

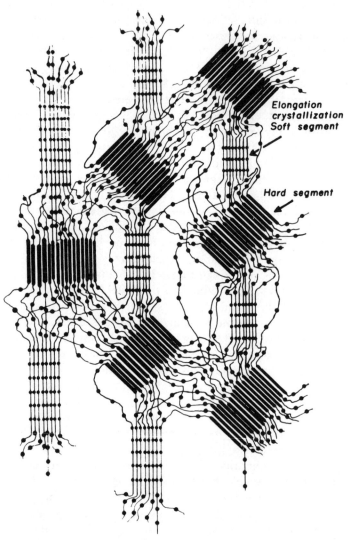

FIG. 18.2. Elongation crystallisation of polyether soft segments in a segmented polyurethane elastomer at 200 per cent extension [14].

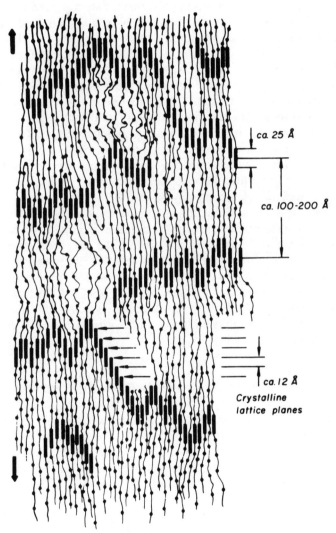

ca. 25 Å

ca. 100-200 Å

ca. 12 Å

Crystalline
lattice planes

FIG. 18.3. Segmented polyurethane elastomer at 500 per cent extension and placed in warm water at 80°C [14].

also evidence [21] for hydrogen-bonded interactions between hard segment NH-groups and soft segment oxygen atoms (*i.e.*, $C:O$ group in polyester and COC-group in polyether). This has been explained in terms of incomplete domain separation resulting in hard segments being dispersed in the soft matrix. It is also postulated that part of these interactions may occur at the domain–matrix interface.

The phenomenon of microphase separation has been clearly demonstrated by electron microscope techniques [29, 30]. Koutsky [29] showed the presence of domains in polyester and polyether-based elastomers by staining samples with iodine and observing darkened areas by transmission electron microscopy. Recent work has shown that the presence of a crystallisable segment in a segmented polyurethane elastomer system, can lead to the development of large-scale structure (termed 'superstructure'). Wilkes *et al.* have demonstrated this phenomenon in materials incapable of hydrogen bonding [19] as well as hydrogen-bonded elastomers [20]. It appears that the superstructure entities ('spherulites') contain preferentially oriented domains. Spherulitic structure of the soft segment is readily apparent at high soft-segment content, whereas at lower soft-segment content, spherulites are formed by aggregation of hard segment domains. Slowikowska and Daniewska [31] reported the formation of radial spherulites in polyurethane elastomer systems, and showed that increased degree of crosslinking in these systems impedes the ability to crystallise.

18.3 STRUCTURE–PROPERTY CORRELATIONS

Thermoplastic elastomers based on various other types of block copolymers appear to derive a high level of physical properties from their domain structure, examples being SBS poly(styrene$_n$-butadiene$_m$-styrene$_n$) and SIS poly(styrene$_n$-isoprene$_m$-styrene$_n$). In these materials, incompatability of the two blocks gives rise to formation of discrete glassy polystyrene domains in a soft rubbery continuous matrix. Obviously, these structures are very similar to those described for polyurethane elastomers, and structure–property relationships are based on the same principles. Direct comparisons of domain structure reinforcement in phase-separated elastomeric block copolymers and filler reinforcement in conventional rubbers have been made by Bishop and Davison [32]. They showed that the reinforcement activity of a polystyrene domain is comparable to that of a good reinforcing carbon black (e.g., HAF black). Morton and Healy [33, 34] investigated the reinforcement characteristics of spherical polystyrene particles dispersed in SBR vulcanisates. Results demonstrated that the tensile strength increases with increased concentration and decreased size of particles. By using different plastics as filler, it was also shown that increased modulus of the filler improves the tensile strength of the filled vulcanisates. Analogously the concentration, size and perfection of hard-segment domains influence the properties of a polyurethane elastomer.

Increased proportion of hard segment in segmented polyurethane elastomers has been shown to give materials of higher modulus [35], an effect observed on increasing the filler content of conventional rubbers. The phenomenon of stress softening on repeated extension is demonstrated in segmented polyurethane elastomers [36] and has been attributed to disruption of domain structure, leading to a decrease in the number of effective 'crosslinking' sites [37, 38]. Smith and Dickie [39] have demonstrated that at a given strain level, the stress in phase separated elastomeric block copolymers decreases as the temperature is raised, owing to melting or softening of domain structures, which therefore become ineffective as tie points and filler particles.

Chang and Wilkes [20] correlated morphological structure in segmented polyurethanes with stress–strain behaviour. They showed that the modulus decreases and extensibility increases with decreasing hard-segment content in poly(ethylene oxide)-based elastomers extended with either ethylenediamine or p-phenylenediamine. The trends were explained in terms of greater degree of domain formation and perfection with increasing hard-segment content. At low hard-segment content (< ~10 per cent), poor domain formation results and crystallisation of the soft segment accounts for the unexpectedly high modulus observed. Comparison of both modulus and breaking stress in ethylenediamine- and p-phenylenediamine-extended materials showed higher values for p-phenylenediamine. This would be predicted as p-phenylenediamine was shown to give greater phase separation and enhanced domain formation.

Estes et al. [40] have also shown an increase in modulus and tensile strength with increasing hard-segment content in a polyether-based material. Measurement of tensile properties over a broad temperature range demonstrated a decrease of stress at a given strain with increasing temperature. The negative temperature coefficient of stress was explained in terms of the visco-elastic softening of hard-segment domains resulting in a decrease in effective physical crosslinks. Smith [41, 42] has reviewed the factors which contribute to strength and toughness in polyurethane elastomers with particular emphasis on the rôle of the dispersed phase. It was pointed out that tear strength is affected in a similar manner to tensile strength.

The following strengthening processes are thought to be effective in a two phase system:

Matrix	Dispersed phase
1. Visco-elastic dissipation of energy near crack tip	1. Increased dissipation of energy
2. Strain-induced crystallisation	2. Deflection and bifurcation of cracks
3. Development of high orientation	3. Induced cavitation
	4. Plastic deformation of domains

Depending on the structure of an elastomer, several or possibly all of these processes may be active. For an elastomer to exhibit retention of strength over an extended range of temperature, it is necessary for a dispersed phase to be present. This will normally either take the form of strain-induced crystalline entities or else the presence of finely divided filler particles. It is thought that domain structures in segmented polyurethane elastomers also act in this latter manner.

At elevated temperatures it is thought [41] that the strength of polyurethane elastomers is dependent primarily on changes in the viscous nature of the matrix and the stress required to induce plastic flow of domain structures. Smith [41] investigated the change in tensile strength over a broad temperature range ($-40°C$ to $+160°C$) for various two-phase materials. Results showed that in the case of triblock elastomers, the tensile strength falls rapidly in a relatively narrow temperature range depending on the softening temperature of domains. Segmented polyurethane elastomers exhibit a rather less pronounced decrease in tensile strength, presumably owing to the presence of more poorly ordered domain structure and a small proportion of covalent crosslinking. Nevertheless, substantial increase in tensile strength with reduction in temperature may be primarily attributed to the progressive increase in domain toughness in polyurethane elastomers. An increase in the viscous characteristics of the matrix is thought to provide a secondary strengthening process.

18.4 CORRELATION OF THE CHEMICAL STRUCTURE OF INDIVIDUAL MORPHOLOGICAL COMPONENTS WITH PHYSICAL PROPERTIES

18.4.1 Soft segment

The flexible (soft) segments in polyurethane elastomers greatly influence the elastic nature of the material and also the properties at low temperature. Aliphatic polyethers and polyesters are the most important materials used to form the soft segments. They have low glass transition temperatures (below room temperature) and are generally amorphous or have low melting points. Polyethers generally give elastomers having a lower level of physical properties than the polyester-based materials, owing to the weaker interchain attractive forces present.

Both polyethylene adipate (PEA) and poly(oxytetramethylene) (POTM) materials crystallise on extension [43] owing to their structural regularity, and this is thought to be an important factor contributing to the high tensile strengths. (POTM gives the best physical properties among polyethers.) The higher elongation at break shown by the polyether material can be explained by the weaker interchain attractive forces present, allowing increased chain slippage and disentanglement. The molecular weight of the soft segment has a marked influence over the final elastomer properties. Increasing the molecular weight relative to the hard segment produces a fall in modulus and an increase in the elongation at break. This is explained by the increase in flexibility and the relative reduction in highly polar hard-segment interactions. Very low molecular weight soft segments (<600) give poorly elastic, hard materials, whereas high molecular weights give soft materials having poorer physical properties.

18.4.2 Hard segment

Interchain attractive forces between rigid segments are far greater than those present in the soft segments, owing to the high concentration of polar groups and the possibility of extensive hydrogen bonding. Hard segments significantly affect mechanical properties,

332   D. C. Harget and C. Hepburn

particularly modulus, hardness and tear strength. The performance of elastomers at elevated temperatures is very dependent on the structure of the hard segments and their ability to remain associated at these temperatures.

DIISOCYANATE CONTRIBUTION

The effect of the diisocyanate structure on the physical properties of polyurethane elastomers has been investigated by several workers [44–47]. Bulky aromatic diisocyanates having a symmetrical molecular structure have been shown to give elastomers of high modulus and hardness and Table 18.1 illustrates this general relationship [47].

TABLE 18.1
Effect of diisocyanate structure on physical properties of segmented polyurethane elastomers [47]

Diisocyanate	Tensile strength (MPa)	Elongation at break (%)	Tear strength (kN/m)	300% Modulus (MPa)	Hardness (°IRHD)
p-PDI[a]	44·1	600	52·5	15·8	72
1,5-NDI[b]	29·4	500	35·3	20·6	80
2,4/2,6-TDI[c]	31·4	600	26·5	2·5	40
MDI[d]	54·4	600	47·1	11·0	61

[a]p-PDI = p-Phenylenediisocyanate; [b]1,5-NDI = 1,5-naphthalene diisocyanate [c]2,4/2,6-TDI = mixed isomers of toluene diisocyanate; [d]MDI = diphenyl methane diisocyanate.

The bulky 1,5-NDI is shown to give materials of higher modulus and hardness than the single aromatic ring p-PDI and the more flexible MDI. Asymmetrical molecules as represented by the 2,4/2,6-TDI combination, give elastomers of low modulus and hardness. Tensile strength and tear strength are also shown to be greater in the case of symmetrical molecules, particularly those based on the 1,4-substituted benzene ring system (p-PDI and MDI).

CHAIN EXTENSION AGENT CONTRIBUTION.

Chain extension agents most commonly employed in polyurethane elastomer systems are diols and diamines. Diols give elastomers having polyurethane hard segments, whereas diamines form essentially polyurea rigid segments. This fundamental structure difference between diol and diamine extended materials generally leads to differences in physical properties between the two classes. Diamine extended materials usually possess a higher level of physical properties owing to the strong hydrogen-bonded interaction of the urea group. This is shown in Table 18.2 for a polyether /TDI system [48].

In the case of diol extended materials, a significant proportion of crosslinking is often introduced by the use of a triol (*e.g.*, trimethylolpropane) to give improved properties. However, these materials are still of low hardness and strength compared with the diamine extended materials (*e.g.*, MOCA extended). Use of amine blends (*e.g.*, MOCA/MDA/*m*-PDA) gives softer materials of lower modulus than those obtained using a single diamine. This is thought to be due to structural irregularity causing decreased intermolecular bonding.

TABLE 18.2
Comparison of amine and polyol extended polyurethane elastomers [48]

Chain extension agent	Tensile strength (MPa)	300% Modulus (MPa)	Elongation at break (%)	Hardness (°IRHD)
MOCA[a]	31·7	12·6	450	90
MOCA/MDA/*m*-PDA (60/20/20)[b,c]	31·7	7·3	450	82
1,4-BD/TMP (1·0/0·3)[d,e]	8·9	2·1	560	57
1,4-BD/TMP (3·0/1·3)	10·4	2·9	470	60

[a] MOCA = 3,3'-Dichloro-4,4'-diaminodiphenylmethane; [b] MDA = methylene dianiline; [c] *m*-PDA = *m*-phenylenediamine; [d] 1,4-BD = 1,4-butanediol; [e] TMP = trimethylolpropane.

From the foregoing discussion the ability of urethane elastomers to retain their high physical strength and elasticity at elevated temperatures seems mostly dependent on the ability of hard-segment domains to remain intact at higher temperatures: the investigation now described was undertaken to examine the consequences of using high-melting chain-extension agents as a means of providing a critical part of these elastomers' domain structure. Such rigid segments by virtue of their theoretical thermal stability could substantially increase the existing engineering service temperatures of urethane elastomers which presently fail, often catastrophically, when certain critical temperatures are reached presumably owing to melting of their hard-domain physical crosslinks.

18.5   EXPERIMENTAL SECTION

18.5.1   Materials
A polyether-based pre-polymer, poly(oxytetramethylene)glycol/4,4'-dicyclohexylmethanediisocyanate (DuPont) was used in the preparation of all elastomers investigated. In order to minimise any possible isocyanate reactions (*e.g.*, self-addition reactions), the prepolymer was stored in a refrigerator until required. 4,4'-Dicyclohexylmethanediisocyanate was used as supplied (DuPont). Solvent, N,N-dimethylformamide (DMF) (Fisons–SLR grade) was dried over molecular sieve (Type 4A) and distilled through a Claisen Column at 80°C under reduced pressure (80 Torr), using a fine bleed of dry nitrogen. The middle fraction was collected and used immediately. When appropriate a stabilised stannous octoate catalyst, *i.e.*, stannous 2-ethylhexoate (designated Nuocure 28, Durham Chemicals) was used as supplied. White spot nitrogen (B.O.C.) was used throughout and further dried by passing through silica gel guard tubes.

PREPARATION OF POLYURETHANE ELASTOMERS
Chain-extension reactions were carried out in dimethylformamide at 90°C under a nitrogen atmosphere. Reactions were followed to completion by an infrared spectroscopic technique [49]. To prepare polyurethanes of higher hard-segment content, a calculated excess of 4,4'-dicyclohexylmethanediisocyanate was added to the pre-polymer solution prior to chain extension. After de-gassing, polymer solutions were cast onto polished glass plates and the solvent allowed to evaporate over 24 hr at 50°C. Resulting polyurethane films were approximately 0·5 mm in thickness and were conditioned for a minimum of two weeks at 20°C and 65 per cent relative humidity before testing.

18.5.2   Polyurethane materials prepared
Table 18.3 summarises the reactions giving solid polyurethane materials. A nomenclature is adopted, whereby the polyurethane is labelled according to the chain-extension agent incorporated. Figures 18.4 and 18.5 illustrate the chemical structures of materials prepared.

MOLECULAR WEIGHT
Values of the number average molecular weight ($\bar{M}_n$) of polyurethane

TABLE 18.3
Polyurethanes prepared by solution polymerisation

Chain-extension agent	Polyurethane nomenclature	Pre-polymer (–NCO)	Reaction time (hr)	Material description
p-Phenylenediamine	p-PDA	4·75	3·5	FE[a]
p-Phenylenediamine	p-PDA (10 NCO)	10	3·5	LFS[b]
p-Phenylenediamine	p-PDA (15 NCO)	15	3·5	LFS
o-Phenylenediamine	o-PDA	4·75	3·0	FE
1,5-Diaminonaphthalene	1,5-DAN	4·75	50·0	FE
1,5-Diaminonaphthalene	1,5-DAN (10 NCO)	10	50·0	LFS
1,5-Diaminonaphthalene	1,5-DAN (15 NCO)	15	50·0	BS[c]
3,6-Diaminoacridine	3,6-DAA	4·75	40·0	FE
3,6-Diaminoacridine	3,6-DAA (10 NCO)	10	40·0	BS
3,6-Diaminoacridine	3,6-DAA (15 NCO)	15	40·0	BS
2,7-Diaminofluorene	2,7-DAF	4·75	10·0	FE
2,3-Diaminofluorene	2,3-DAF	4·75	3·0	FE
2,4-Diamino-6-phenyl-s-triazine	2,4-DPT[d]	4·75	30·0	FE
3,3'-Dichloro-4,4'-diaminodiphenylmethane	MOCA[d]	4·75	15·0	FE
1,4-Butanediol	1,4-BD[d]	4·75	8·0	FE

[a] FE = Flexible elastomer; [b] LFS = low flexibility solid; [c] BS = brittle solid; [d] catalysed with 1 per cent stannous octoate.

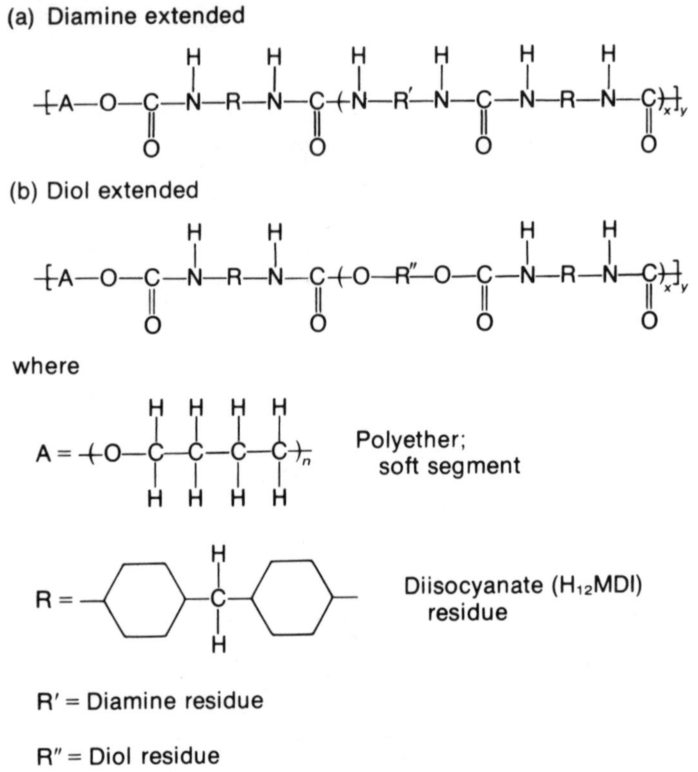

(a) Diamine extended

(b) Diol extended

where

$$A = \{O-\underset{\underset{H}{|}}{\overset{\overset{H}{|}}{C}}-\underset{\underset{H}{|}}{\overset{\overset{H}{|}}{C}}-\underset{\underset{H}{|}}{\overset{\overset{H}{|}}{C}}-\underset{\underset{H}{|}}{\overset{\overset{H}{|}}{C}}\}_n$$ Polyether; soft segment

R = Diisocyanate (H_{12}MDI) residue

R' = Diamine residue

R'' = Diol residue

FIG. 18.4. Chemical structure of polyurethane elastomers.

materials investigated are given in Table 18.4. Values were deter
mined by membrane osmometry except where stated. The majority o
$\bar{M}_n$ values lie in the region of 20 000–30 000. MOCA, 2,4-DPT, 3,6
DAA (10 NCO) and 3,6-DAA (15 NCO) are of significantly lowe
molecular weight and this factor should be taken into account whe
comparisons of physical property data are made. It is suggested tha
these relatively low values of $\bar{M}_n$ are due to low molecular weigh
fractions resulting from a predominance of amine 'capping' reaction
over amine 'extension' reactions.

SOLUBILITY
The majority of the polyurethane materials were found to be soluble i
DMF after shaking for 2 hr at room temperature. 1,5-DAN an

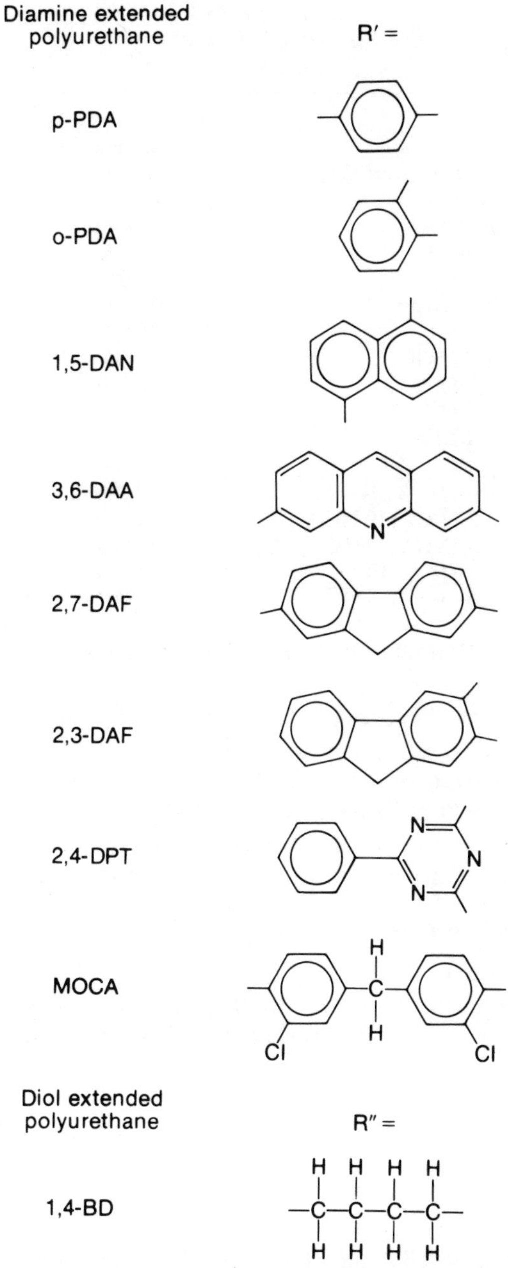
Diamine extended polyurethane
R' =
p-PDA
o-PDA
1,5-DAN
3,6-DAA
N
2,7-DAF
2,3-DAF
2,4-DPT
N
N
N
MOCA
H
C
H
Cl
Cl
Diol extended polyurethane
R'' =
1,4-BD
H H H H
C C C C
H H H H

FIG. 18.5. Chain-extension agents investigated.

TABLE 18.4
*Number average molecular weights
of polyurethane elastomers*

Polyurethane	Number average molecular weight ($\bar{M}_n$)
p-PDA	28 000
o-PDA	19 400
1,5-DAN	24 000
3,6-DAA	22 000
2,7-DAF	30 000
2,3-DAF	19 000
2,4-DPT	12 000[a]
MOCA	11 500[a]
1,4-BD	20 000
p-PDA (10 NCO)	18 000
p-PDA (15 NCO)	19 000
1,5-DAN (10 NCO)	22 000
1,5-DAN (15 NCO)	19 000
3,6-DAA (10 NCO)	13 000[a]
3,6-DAA (15 NCO)	12 000[a]

[a]Determined by vapour pressure osmometry.

2,7-DAF required heating to 60°C and were found to dissolve completely at this temperature. 1,5-DAN (10 NCO) and 1,5-DAN (15 NCO) dissolved on heating, leaving a small amount of gel phase which was thought to be due to a fraction of crosslinked material. The solubility of the polyurethane elastomers provides evidence of the essentially linear molecular structure of these materials.

18.5.3 Tensile properties

TENSILE STRENGTH
The results of tensile properties determined over the temperature range −25°C to +150°C are presented in Figs. 18.6 and 18.7. Although it would not be meaningful to attempt to correlate relatively minor variations in properties shown by different chain extension agents, it is worthwhile observing trends arising from this data. (See Authors' Note, p. 357.)

MODULUS
Figures 18.8 and 18.9 illustrate the effect of each of the above variables on 100 per cent modulus values over the temperature range

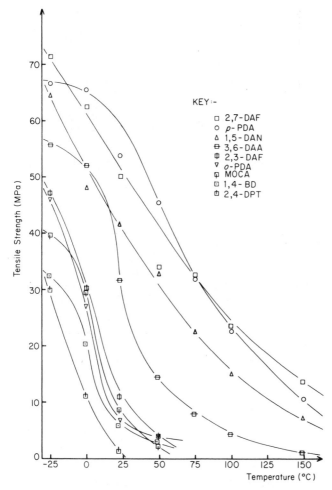

KEY:-

□ 2,7-DAF
○ *p*-PDA
△ 1,5-DAN
⊖ 3,6-DAA
⊞ 2,3-DAF
▽ *o*-PDA
⊡ MOCA
▫ 1,4-BD
⊕ 2,4-DPT

FIG. 18.6. Ultimate tensile strength of polyurethane elastomers, −25°C to +150°C.

25°C to +150°C. In general it can be seen that the same trends arise
as were observed for tensile strength. A notable effect of hard-
segment content is illustrated in Fig. 18.9. Modulus values increase
with increased hard-segment content at all temperatures. It would
appear that crystallisation of the soft segment plays only a minor rôle
at 100 per cent elongation and that the physical association of hard
segments is the important reinforcing factor. At temperatures below
the melting point of crystalline POTM (43°C), modulus is therefore

D. C. Harget and C. Hepburn

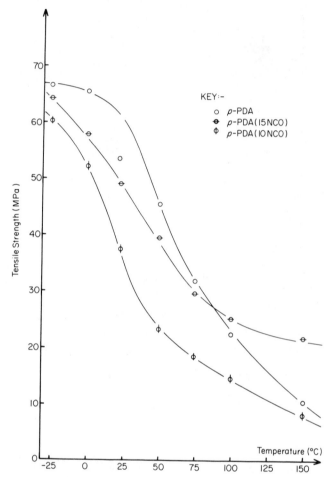

FIG. 18.7. Ultimate tensile strength of *p*-PDA polyurethanes. Effect of pre-polymer NCO content, −25°C to +150°C.

proportional to hard-segment content. At higher temperatures, this relationship still holds true, as the thermal stability of the hard-segment aggregates is greater for materials of higher hard-segment content.

ELONGATION AT BREAK

The effect of hard-segment structure on 'elongation at break' (EB) is

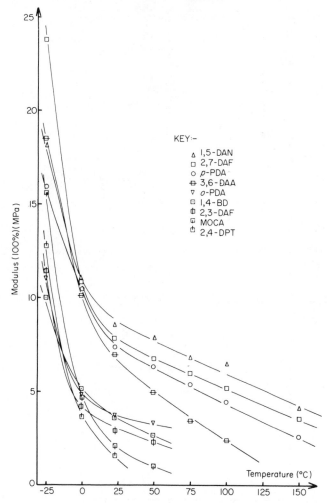

FIG. 18.8. Modulus (100 per cent) of polyurethane elastomers, −25°C to +150°C.

llustrated in Figs. 18.10 and 18.11. From −25°C all materials show
ncreased EB with increased temperature. Asymmetric groups exhibit
igher elongation initially but fail at relatively low temperatures, *e.g.*,
⁕-PDA, 2,3-DAF, 2,4-DPT. More symmetrical groups give lower
nitial elongation, but continue to increase with increasing tem-
⁕erature, *e.g.*, *p*-PDA, 2,7-DAF, 1,5-DAN. This effect may be

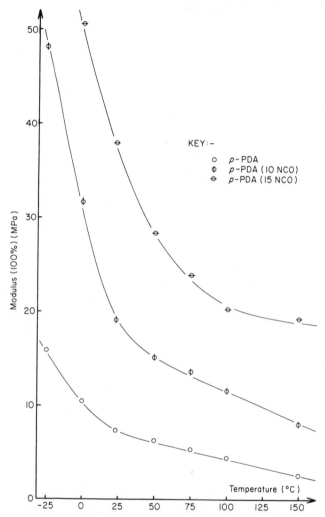

FIG. 18.9. Modulus (100 per cent) of *p*-PDA polyurethane. Effect of pre-polymer NCO content, −25°C to +150°C.

explained by the poor interchain attraction available in the asymmetric groups, leading to relatively unrestricted disentanglement and flow of polymer chains. With increased temperature, essentially no 'tie points' are available and materials fail at low extension. On the

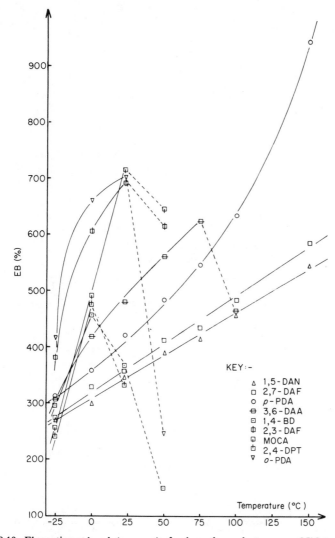

FIG. 18.10. Elongation at break (per cent) of polyurethane elastomers, −25°C to +150°C.

other hand, symmetrical groups give rise to strong interchain attractive forces which provide strong 'tie points' through hard-segment
aggregation thus leading to restricted chain disentanglement and poor

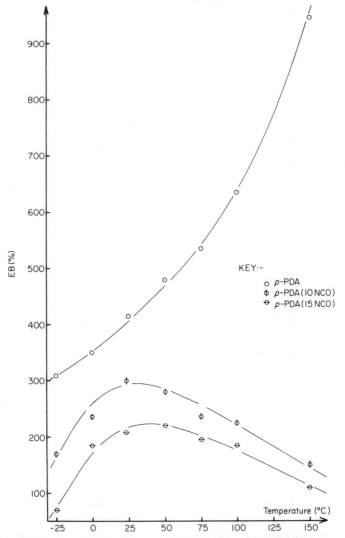

FIG. 18.11. Elongation at break (per cent) of p-PDA polyurethanes. Effect of pre-polymer
NCO content, $-25°C$ to $+150°C$.

slippage of polymer chains past each other. As the temperature is
increased towards 150°C, the effectiveness of these 'tie points'
decreases, and EB increases.

Figure 18.11 illustrates the effect of hard-segment content variation on EB. Over the entire temperature range, EB increases with decreasing hard-segment content. This may be explained by the higher proportion of intermolecular attractive forces available at higher NCO levels, thus giving rise to strong 'tie points' which remain effective, even at 150°C.

TEAR

The tear propagation strength, T_p (*i.e.*, the force per unit thickness required to propagate a tear through the material) and the total tear energy, T_e (*i.e.*, the work per unit thickness required to propagate a tear through the material) are shown in Tables 18.5 and 18.6. Materi-

TABLE 18.5
Tear properties of low extensibility
polyurethane elastomers

Polyurethane	Tear strength (propagation), T_p (kN·m^{-1})	Tear energy T_e(J·m^{-1})
p-PDA	12·4	398
1,5-DAN	5·8	165
2,7-DAF	10·2	306
p-PDA (10 NCO)	28·4	825
1,5-DAN (10 NCO)	18·8	498
3,6-DAA (10 NCO)	8·2	220
1,4-BD	9·0	348

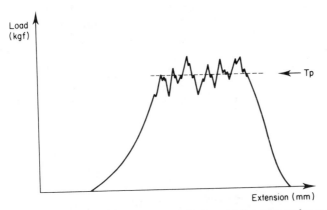

FIG. 18.12. Low extensibility tear propagation curve. T_e = area under curve.

als were found to fall into two distinct types.

Low extensibility (Table 18.5). Elastomers of this type yielded load–extension curves of the type shown in Fig. 18.12. The force required to propagate a tear through the material remains essentially constant and extension of the polymer network is minimal.

High extensibility (Table 18.6). Figure 18.13 illustrates a typical load–extension curve for this type of material. In this case, the polymer network is subject to increasing extension throughout the test period. The tear propagation force therefore increases, and it is

TABLE 18.6
*Tear properties of high extensibility
polyurethane elastomers*

Polyurethane	Tear strength (propagation), $(kN \cdot m^{-1})$		Tear energy T_e $(J \cdot m^{-1})$
	T_p(min.)	T_p(max.)	
3,6-DAA	24·7	39·8	2678
o-PDA	14·7	20·0	1037
2,3-DAF	22·6	24·5	3072
MOCA	11·5	13·3	1450
2,4-DPT	7·0	10·1	630

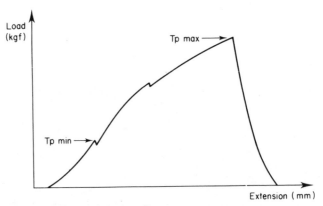

FIG. 18.13. High extensibility tear propagation curve. T_e = area under curve.

convenient to record a minimum (T_p min.) and a maximum ($T_{p\,max.}$) value of tear strength.

Generally it is found that symmetrical groups in the hard segment give rise to low extensibility materials, e.g., p-PDA, 1,5-DAN, 2,7-DAF. Materials with higher hard-segment content also fall into this class, i.e., p-PDA (10 NCO), 1,5-DAN (10 NCO) and 3,6-DAA (10 NCO).

Asymmetrical groups tend to give high extensibility tear, as shown by o-PDA, 2,3-DAF, 2,4-DPT. These general trends are supported by the previous results reported for EB at room temperature. Values of T_p and particularly T_e, are greater for highly extensible materials owing to the additional force and work required to extend the complete polymer network. A comparison of tear values between materials of low and high extensibility is therefore meaningless. However, it is worthwhile comparing values of T_p and T_e within each class.

In general, the more bulky symmetrical groups appear to give lower tear strength and tearing energy than less bulky groups in low extensibility tear. This is illustrated by p-PDA (10 NCO), 1,5-DAN (10 NCO) and 3,6-DAA (10 NCO). p-PDA also gives higher values of T_p and T_e than 1,5-DAN and 2,7-DAF. High tear strength and energy depend primarily on the presence of strong tie points throughout the polymer matrix. These tie points hinder the progress of crack growth and in polyurethanes may take the form of hard-segment domains. It would appear that although the extent of physical crosslinking between hard segments should be similar for the symmetrical groups investigated, the more bulky groups yield a less closely packed domain structure with weaker resistance to crack growth.

Materials having higher hard-segment content are found to give higher values of T_p and T_e. This is demonstrated by comparing p-PDA and 1,5-DAN with p-PDA (10 NCO) and 1,5-DAN (10 NCO). Presumably this is due to the presence of larger and more perfectly developed domains which are more effective as crack growth inhibitors. From Table 18.6 there appears to be little correlation between structure of the chain extender and values of tear strength or tear energy in high extensibility tear.

18.5.4 Stress relaxation

Stress relaxation was measured graphically using the Wallace stress

348 D. C. Harget and C. Hepburn

relaxometer which determines the decrease in force with time with
respect to a specific test sample strain. For comparison of data of a
series of samples it is more meaningful to consider the variation in
relative force, *i.e.*, f(t)/f(0), with time. In this case f(t) and f(0)
represent the force applied by the sample at time '*t*' and at the start of

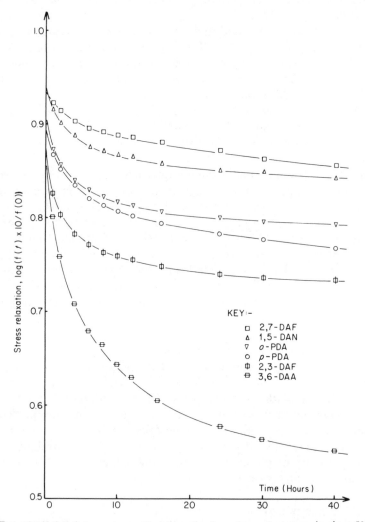

FIG. 18.14. Continuous stress relaxation of polyurethane elastomers in air at 50°C.

the relaxation (*i.e.*, immediately after extension to 50 per cent) respectively. Results are presented graphically as log [(f(*t*) × 10)/f(0)] versus time.

Figures 18.14 and 18.15 show a selection of stress relaxation data for different chain extension agents at fixed temperatures over the

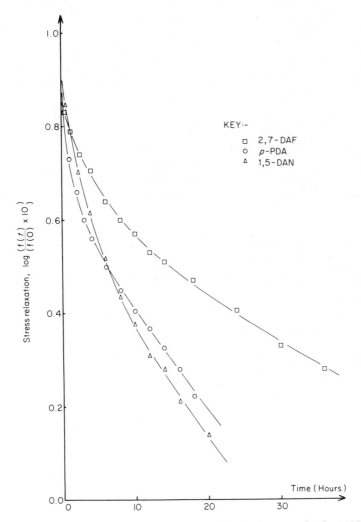

KEY:-

☐ 2,7-DAF
○ *p*-PDA
△ 1,5-DAN

Stress relaxation, log $\left(\frac{f(t)}{f(0)} \times 10\right)$

Time (Hours)

FIG. 18.15. Continuous stress relaxation of polyurethane elastomers in air at 125°C.

range 50°C–125°C. Figure 18.16 illustrates the typical effect of temperature on stress relaxation of the polyurethane elastomers investigated.

Results of measurements made in air were compared with stress relaxation under a nitrogen atmosphere and typical results are shown

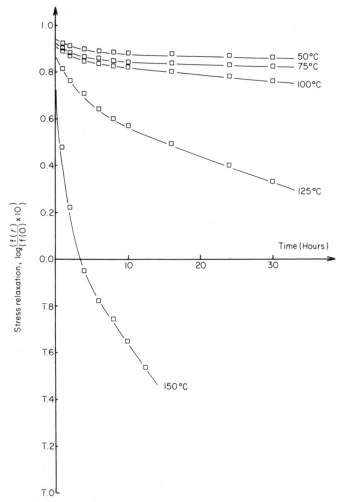

FIG. 18.16. Continuous stress relaxation of 2,7-DAF elastomers in air at different temperatures (50–150°C).

in Fig. 18.17 using 1,5-DAN as an example. 'Intermittent' mode stress relaxation data are presented in comparison with 'continuous' data for *p*-PDA in Fig. 18.18.

From Fig. 18.14 it can be seen that bulky symmetrical groups contribute significantly to low stress relaxation at elevated tem-

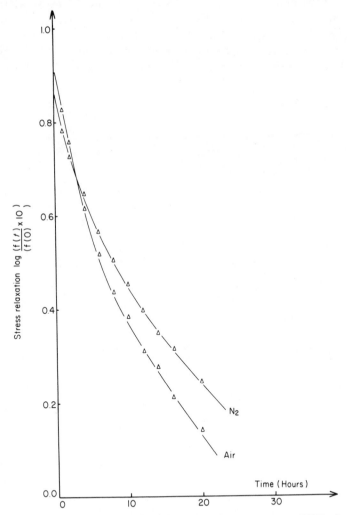

FIG. 18.17. Continuous stress relaxation of 1,5-DAN elastomer at 125°C; air versus nitrogen atmosphere.

perature. Other data, not illustrated, show that between 50°C and 100°C, 1,5-DAN and 2,7-DAF show relatively little relaxation. Above 100°C, both these materials show a rapid increase in stress relaxation (Fig. 18.15) although the more bulky 2,7-DAF shows retention of some stress at 125°C after approximately 30 hr (Fig. 18.16). Results also indicate a pronounced effect of highly asymmetric groupings in polyurethane materials. At 50°C 2,3-DAF and 3,6-DAA show relatively high stress relaxation. This effect is enhanced at higher temperatures.

Figure 18.17 illustrates the effect of an inert atmosphere on stress relaxation. It is seen that only a slight increase in relaxation rate is found in an atmosphere of air compared with a nitrogen atmosphere.

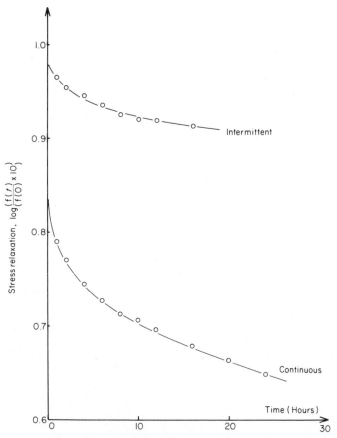

FIG. 18.18. Stress relaxation of p-PDA elastomer in air at 100°C; continuous versus intermittent.

This effect was found in each case investigated suggesting only a limited extent of oxidative cleavage. Unlike the materials investigated by Singh et al. [50], which consisted of highly crosslinked networks, the only covalent crosslinking present in these materials would be a small proportion of allophanate and biuret sites. This difference in structure would account for the apparent contradiction between the present data and those recorded by Singh et al. for polyether-based urethanes. Assuming oxidative cleavage to predominate at crosslink sites, this would account for the minimal extent of oxidative cleavage observed in the materials investigated here.

A comparison of 'intermittent' and 'continuous' stress relaxation data, shown typically in Fig. 18.18, indicates an essentially reversible relaxation process in the materials investigated. Once again, results appear to contradict the data of Singh who demonstrated irreversible relaxation in polyether networks. It is suggested that fundamental structural differences are again responsible for this apparent contradiction. The materials investigated show only a very low extent of relaxation under 'intermittent' conditions, indicating minimal irreversible relaxation. This irreversible portion of stress relaxation is most probably due to chain scission at weak allophanate and biuret crosslink sites.

It is suggested that reversible stress relaxation in segmented linear polyurethanes may be explained by physical relaxation processes which depend greatly on the reorganisation of hard-segment interactions. Physical relaxation processes associated with unrestrained flexible linear soft segments would also be expected to contribute to the total relaxation. These latter processes may include simple chain uncoiling in the direction of applied stress and disentanglement of chains to relieve stress. However, it would be anticipated that such processes are generally considerably restrained by the presence of hard-segment 'tie points' throughout the soft-segment matrix. It can be seen therefore that physical relaxation processes are controlled by the degree of organisation of hard segments. This would explain the observed decrease in stress relaxation with increased symmetry in hard-segment structure. The effect of temperature is also explained. With increasing temperature hard-segment interactions are progressively reduced and dissociation of hard-segment domain structures becomes more pronounced.

Figure 18.19 illustrates three possible relaxation processes which involve dissociation of hard-segment interactions such as hydrogen-bond breakdown. In each case, it can be seen that the effective chain

(1) Dissociation of hard-segment interactions under stress at
     elevated temperature.

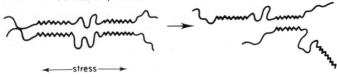

←——stress——→

(2) Intramolecular hydrogen-bond disruption.

←——stress——→

(3) Promotion of chain disentanglement as a result of dissociation
     of hard-segment 'tie points'

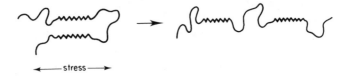

←—— stress ——→

where, ⋀⋀⋀⋀⋀   = hard segment
         ∿∿∿   = soft segment

FIG. 18.19. Physical relaxation phenomena in a linear segmented polyurethane network.

length in the direction of stress is increased, thus reducing the overall
internal stress at constant elongation. On relief of localised stress at a
particular site it may be possible for new hard-segment interactions to
form, as illustrated in process (3) of Fig. 18.19. Here, interaction
between hard segments A and B has dissociated owing to stress at
elevated temperature and a new interaction between A and C has
formed allowing a disentanglement of polymer chains and hence a
relief of localised stress.

In conclusion, data have been presented that demonstrate how the strength and time-dependent properties of polyurethane elastomers can be significantly controlled over a broad temperature range by choice of chain-extension agent. Symmetrical high-melting chain-extension agents as typified by p-PDA, 1,5-DAN, and 2,7-DAF exhibit better retention of properties at higher temperatures than do their asymmetrical analogues.

Polyurethanes were impressed upon Bob Payne early in his SATRA career as important new shoe making materials requiring radically different shoe making processes. These aroused his keen interest and at his initiative many investigations were undertaken to better understand their full capabilities and limitations in the footwear field to the extent that now they are established there on a sound basis of science and technology.

REFERENCES

1. J. H. SAUNDERS and K. C. FRISCH. (1962). *Polyurethanes: Chemistry and Technology–I Chemistry*, Ch. 2. New York & London: Interscience.
2. S. L. AGGARWAL. *Polymer*, 1976, **17**, 938.
3. D. C. ALLPORT and A. A. MOHAJER. (1973). *Block Copolymers* (Eds D. C. ALLPORT and W. JANES), Ch. 5. London: Applied Science Publishers.
4. G. W. MILLER and J. H. SAUNDERS. *J. Appl. Poly. Sci.*, 1969, **13**, 1277.
5. G. W. MILLER and J. H. SAUNDERS. *J. Poly. Sci.*, 1970, **A1(8)**, 1923.
6. H. N. NG, A. E. ALLEGREZA, R. W. SEYMOUR and S. L. COOPER. *Polymer*, 1973, **14**, 255.
7. N. S. SCHNEIDER, C. S. PAIK SUNG, R. W. MATTON and J. L. ILLINGER. *Macromolecules*, 1975, **8(1)**, 62.
8. G. W. MILLER. *J. Appl. Poly. Sci.*, 1971, **15**, 39.
9. S. B. CLOUGH and N. S. SCHNEIDER. *J. Macromol. Sci.*, 1968, **B(2)**, 553.
10. D. S. HUH and S. L. COOPER. *Poly. Eng. Sci.*, 1971, **11**, 369.
11. R. W. SEYMOUR and S. L. COOPER. *J. Poly. Sci.*, 1971, **B(9)**, 689.
12. R. W. SEYMOUR and S. L. COOPER. *Macromolecules*, 1973, **6**, 48.
13. N. S. SCHNEIDER and C. S. PAIK SUNG. *Poly. Eng. Sci.*, 1977, **17(2)**, 73.
14. R. BONART. *J. Macromol. Sci.*, *Physics*, 1968, **B(2)**, 115.
15. R. BONART, L. MORBITZER and G. HENTZE. *J. Macromol. Sci.*, *Physics*, 1969, **B(3)**, 337.
16. R. BONART, L. MORBITZER and E. MULLER. *J. Macromol. Sci.*, 1974, **(B)9**, 447.
17. R. BONART and E. MULLER. *J. Macromol. Sci.*, 1975, **(B)10**, 177, 345.
18. S. B. CLOUGH, N. S. SCHNEIDER and A. O. KING. *J. Macromol. Sci.*, *Physics*, 1968, **(B)2**, 641.
19. S. L. SAMUELS and G. L. WILKES. *J. Poly. Sci.*, 1973, **(C)43**, 149.
20. Y. P. CHANG and G. L. WILKES. *J. Poly. Sci.*, *Physics*, 1975, **13**, 455.

21. R. SEYMOUR, G. M. ESTES and S. L. COOPER. *Poly. Preprints*, 1970, (C)11, 867.
22. C. S. PAIK SUNG and N. S. SCHNEIDER. *Macromolecules*, 1975, 9(1), 68.
23. Y. M. BOYARCHUK, L. RAPPOPORT, V. NIKITIN and N. APUKHTINA. *Poly. Sci. USSR*, 1965, 7, 859.
24. T. TANAKA and T. YOKOYAMA. *J. Poly. Sci.*, 1968, (C)23, 865.
25. K. NAKAYAMA, T. INO and I. MITSUBARA. *J. Macromol. Sci., Chem.*, 1969, (A)3, 1005.
26. T. TANAKA, T. YOKOYAMA and Y. YAMAGUCHI. *J. Poly. Sci.*, 1968, (A1)6, 2137.
27. T. TANAKA, T. YOKOYAMA and Y. YAMAGUCHI. *J. Poly. Sci.*, 1968, (A1)6, 2153.
28. H. ISHIHARA *et al. J. Macromol. Sci., Physics*, 1974, (B)10, 591.
29. J. KOUTSKY. *Poly. Letters*, 1970, 8, 353.
30. R. R. LAGASSE. *J. Appl. Poly. Sci.*, 1977, 21, 2489.
31. I. SLOWIKOWSKA and I. DANIEWSKA. *J. Poly. Sci., Poly. Symp.*, 1975, 53, 187.
32. E. T. BISHOP and S. DAVISON, *J. Poly. Sci., Part C, Poly. Symp.*, 1969, 26, 59.
33. M. MORTON. (1971). *Advances in Chemistry—Vol. 99* (Ed. R. F. GOULD), p. 490. Washington: American Chemical Society.
34. M. MORTON and J. C. HEALY. *Poly. Preprints*, 1967, 8, 1569.
35. S. L. COOPER and A. V. TOBOLSKY. *Rubb. Chem. Tech.*, 1967, 40, 1105.
36. D. PUETT. *J. Poly. Sci.*, 1967, (A2)5, 839.
37. J. F. BEECHER, L. MARKER, R. D. BRADFORD and S. L. AGGARWAL. *J. Poly. Sci.*, 1969, (C)26, 117.
38. E. FISCHER and J. F. HENDERSON. *J. Poly. Sci.*, 1969, (C)26, 149.
39. T. L. SMITH and R. A. DICKIE. *J. Poly. Sci.*, 1969, (C)26, 163.
40. G. M. ESTES, D. S. HUH and S. L. COOPER. (1970). *Block Polymers* (Ed. S. L. AGGARWAL), p. 225. New York: Plenum.
41. T. L. SMITH. *J. Poly. Sci., Physics*, 1974, 12, 1825.
42. T. L. SMITH. *Poly. Eng. Sci.*, 1977, 17(3), 129.
43. L. WEISFELD, R. J. LITTLE and W. E. WOLSTENHOLME. *J. Poly. Sci.*, 1962, 56, 455.
44. G. TRAPPE. (1968). *Advances in Polyurethane Technology* (Eds J. M. BUIST and H. GUDGEON), Ch. 3. London: Maclaren.
45. C. S. SCHOLLENBERGER. (1969). *Polyurethane Technology* (Ed. P. BRUINS), Ch. 10. New York: Interscience.
46. J. H. SAUNDERS. *Rubb. Chem. Tech.*, 1960, 33(5), 1259.
47. K. PIGOTT, B. F. FRYE, K. R. ALLEN, S. STEINGISER, W. C. DARR and J. H. SAUNDERS. *J. Chem. Eng. Data*, 1960, 5(3), 391.
48. R. J. ATHEY, J. G. DIPINTO and J. S. RUGG. *Adiprene L, A Liquid Urethane Elastomer.* Dev. Prod. Rep. No. 10, March 1958, DuPont.
49. D. C. HARGET. Ph.D. Thesis. (1977). Loughborough University of Technology.
50. A. SINGH, L. WEISSBEIN and J. C. MOLLICA. *Rubb. Age*, 1966, 98(12), 77.

AUTHORS' NOTE

Figures 18.6 and 18.7 show variation of ultimate tensile strength with temperature. From Fig. 18.6 it can be seen that materials studied fall into two basic groups. 2,7-DAF, p-PDA and 1,5-DAN all have high tensile strengths at ambient temperature. With increasing temperature, tensile strength falls off, although a significant proportion is retained even at 150°C. o-PDA, 2,3-DAF, 2,4-DPT, 1,4-BD and MOCA exhibit relatively low tensile strength at ambient temperature, with little retention of strength at temperatures of 50°C and above. 3,6-DAA appears to give properties which fall between these two extremes. The first group consists of polymers having highly symmetrical groupings in the hard segment. These groupings apparently lead to ordered domain structure resulting in a high level of physical properties. At 150°C domain structure is still not completely disrupted and hence a significant strength is retained. The second group consists essentially of polymers incorporating asymmetrical hard segment units, i.e. o-PDA, 2,3-DAF and 2,4-DPT. Poor ordering results in these systems yielding polymers of low room temperature strength and almost total loss of strength at elevated temperatures. The poor tensile strength of MOCA is thought to be due to its relatively low molecular weight (Table 18.4). This may also partly account for the low strength of 2,4-DPT.

3,6-DAA shows a level of tensile strength below that of the symmetrical materials over the whole temperature range. This is reasonable, as the bulky three ring unit presents a certain amount of steric hindrance between adjacent hard segments due to the 3,6 substitution pattern. It may be anticipated that 1,4-BD would yield good hard segment domain formation due to the regular structural unit involved. However, the low level of tensile strength properties exhibited by this material indicates the presence of a disordered system having mixed hard and soft segments. It is suggested that mixing is promoted by the common —$(CH_2)_4$—O— group in both hard and soft segments.

Figure 18.7 illustrates the effect of hard segment content on tensile strength for p-PDA. At ambient temperature p-PDA elastomers exhibit significantly higher tensile strength than materials based on higher NCO systems. This is explained by the higher elongation achieved in p-PDA and the much greater ratio of soft segment to hard segment. Both these factors give rise to the relative ease of soft-segment crystallisation in p-PDA which results in a particularly high tensile strength. With increasing temperature, retention of tensile strength becomes higher in the case of

p-PDA (10 NCO) and p-PDA (15 NCO), owing to the greater reinforcement afforded by the strong domain structures derived from the long hard segments present.

AUTHOR INDEX

SUBJECT INDEX